GRAMMAR, **GEOMETRY,** & *Brain*

CSLI LECTURE NOTES NUMBER 200

GRAMMAR,
GEOMETRY,
& *Brain*

JENS ERIK FENSTAD

CSLI
PUBLICATIONS
Center for the Study of
Language and Information
Stanford, California

Copyright © 2010
CSLI Publications
Center for the Study of Language and Information
Leland Stanford Junior University
Printed in the United States
14 13 12 11 10   1 2 3 4 5

Library of Congress Cataloging-in-Publication Data

Fenstad, Jens Erik.
  Grammar, geometry, and brain / Jens Erik Fenstad.

    p. cm. – (CSLI lecture notes ; no. 200)

  Includes bibliographical references and index.

  ISBN 1-57586-593-9 (cloth : alk. paper) —
  ISBN 1-57586-592-0 (pbk. : alk. paper)
  1. Language and logic.   I. Title. II. Series.

  P39.F46 2009
  401′.9–dc22

                                                            2009032030
                        CIP

∞ The acid-free paper used in this book meets the minimum requirements
of the American National Standard for Information Sciences—Permanence
    of Paper for Printed Library Materials, ANSI Z39.48-1984.

CSLI was founded in 1983 by researchers from Stanford University, SRI
  International, and Xerox PARC to further the research and development of
integrated theories of language, information, and computation. CSLI headquarters
  and CSLI Publications are located on the campus of Stanford University.

CSLI Publications reports new developments in the study of language,
    information, and computation. Please visit our web site at
                http://cslipublications.stanford.edu/
      for comments on this and other titles, as well as for changes
            and corrections by the author and publisher.

Contents

Preface

The aim of this book is to explore the relationship between language, meaning and brain. I would have liked to tell one connected story, moving from grammar and logic, through meaning and mind, toward an attempt to link language and brain. There is no lack of texts trying to do precisely this. But almost all of such texts have had to resort to some wishful thinking in putting all the pieces together in one cohernt account. I do not claim to present a final solution; we still have to live with a plurality of approaches. My strategy has been to gather several "hard" pieces of the relevant sciences, hoping that they in some form will be part of a more comprehensive story. Science is shaped by ignorance, and the art of science consists in doing what is do- able. I hope that I, in addition to presenting some facts and theories, have been able to point to further possibilities and new challenges, well within the range of what is do-able.

For whom is the book written? I hope to reach both the linguistic and cognitive science communities. For the latter community, including, in particular, the cognitive neuroscientists, I hope that the earlier chapters of the book will show what kind of higher order structures brain dynamics will have to explain. To the linguists and philosophers I hope to demonstrate in the last part of the book that grounding cognitive behavior in brain dynamics requires more than current neural network simulations.

In addition, I also have the hope that someone from the next generation, well-trained in the mathematical sciences, will read the text and be encouraged to apply their particular skills to the complex problems of language, meaning and brain. The book is not a text-book, but could, perhaps, serve as a useful guide.

What will be the background needed to read the book? This is a difficult question. There is obviously a risk of writing a book that is comprehensible only to the author. I hope not. The reader will, however, need to possess sufficient mathematical "culture", in particular to understand the power and limits of mathematical modeling. The text is not a "mathematical text", but the mathematics is always there as part of the sub-text. And at certain points it needs to be made explicit.

Acknowledgments. In 1976 I was invited to participate in a seminar at the University of Oslo on the foundation of language organized by the linguist Even Hovdhaugen, the psychologist Ragnar Rommetveit, and the philosopher Dagfinn Føllesdal. The topic was fashionable and we had an overflow audience, at least until the hired logician started to lecture on the technicalities of higher order intensional logic and the associated grammar formalisms. Not everbody, however, disappeared; the seminar continued for almost 20 years and was a useful link between logic and linguistics at the University of Oslo. Many linguists participated. I want, in particular to mention Catrine Fabricius-Hansen and Kjell Johan Sæbø. I was also able to form a research group in logic and natural language systems at the Department of Mathematics. The topic turned out to be popular, and we had many students. We shared a common enthusiasm for the subject and I want to thank them all, first as students, later as postdocs and younger colleagues, for their contributions to the project: Jan Tore Lønning, Tore Langholm, Helle Frisak Sem, Erik Colban, Espen Vestre, and Guri Værne. Much of what I report on in this book represents joint work with them.

I spent the year 1983-84 at Stanford University. This was the founding year of CSLI, the Center for the Study of Language and Information, and there was much activity across disciplinary boundaries. I learned much from many people and want to thank, in particular, Jon Barwise, Pat Suppes and Stanley Peters for sharing their knowledge and insights with me. But above all I want to thank my co-authors of the book *Situations, Language and Logic*, which we wrote during that year: Per-Kristian Halvorsen, Tore Langholm and Johan van Benthem.

The cooperation with CSLI and XEROX continued over the years, e.g. much of chapter 2 was written at XEROX during a stay there as visiting scientist in 1994. But there have also been other important contacts with linguists, logicians and cognitive scientists on the themes of this book. Of great importance has been the close link with Peter Gärdenfors, logician and cognitive scientist at the University of Lund. His work on conceptual spaces has been a decisive influence on my own work on formal semantics and geometric model theory, as reported in chapter 2. I would also like to mention John Goldsmith, linguist at the University of Chicago, for many helpful comments, particularly on the themes of chapter 1. Dagfinn Føllesdal, philosopher at Oslo and Stanford, was mentioned above as organizer of the seminar in 1976. But his influence extends far beyond the role as organizer. We have been friends and colleagues since the late 1960s, when he returned from Harvard and Quine and I returned from Berkeley and Tarski. I have benefitted from his good advice and critical remarks on much of what I have written, including this text. I also want to thank Dag Normann for his invaluable help on the final editing of the text.

Finally, sincere thanks to my wife Grete, who has given invaluable advice on language and style, as well as taken care of the complexities of the LaTeX system in the writing of this text.

Introduction

Brain and cognition is the great challenge of science today. A precise understanding of language and meaning and how they emerge out of brain structure are central parts of this challenge. The study of language and meaning must necessarily proceed on many levels. At the "top" level we need a theory of grammar with equal emphasis on the combinatorial and the conceptual parts. At the "bottom" level we need a precise understanding of the structure and functioning of the brain. And the two must be connected.

We have in recent years seen great advances in linguistics and the related cognitive sciences. Experimental and observational techniques at all levels from basic neuroscience to the study of grammar and text have been transformed and lifted to new levels of sophistication. Modeling skills and theory have been significantly expanded using increased insights from the mathematical and natural sciences. Simulations have been extensively used to mimic cognitive behavior. But we do not yet fully understand what human beings so effortlessly can do, how to connect speech, symbol and meaning into one seamless unit.

There are no lack of general accounts and of strong claims of having found the "final solution" to this challenge. And, indeed, beyond well confirmed observations and generally accepted facts there are many plausible stories to tell. But almost always they remain just that, stories and not "hard" science.

I have not found the "final solution". The aim of this account is to review some aspects of what has been achieved so far. And being a mathematician I have whenever possible selected examples of "hard" models to move the story forward. The text is organized in three main parts.

Grammar and logic

The relationship between logic and language has always been a complex affair. They have been inseparably intertwined in the European intellectual tradition. It was not an easy relationship, but there were always issues of substance whether they were friends or foes. I shall in the first chapter tell parts of the history of how logic and grammar have interacted from ancient to modern times. In addition we shall briefly review a few high points of this interaction as it is seen today, thus providing a proper foundation for our inquiry into the structure of grammar and how it relates to meaning and brain.

It is claimed that Chomsky revolutionized the study of language. In many ways this is true, but there could have been an alternative story to tell. In fact, there was a second, but largely hidden, revolution. In the 1930s and 1940s we saw a number of fundamental contributions to logic and grammar, highlighting the links between categorial grammar and higher-order logic, and associated with names such as K. Ajdukiewicz, H. Reichenbach and H. B. Curry. But neither linguists nor logicians paid any attention. It seems that the dominating philosophical doctrines of the time forbade such "metaphysical" speculations.

The Chomsky revolution also holds another lesson: You are never more than the mathematics you know. At the time of *Syntactic Structures* model theory and formal semantics were disciplines still in an emergent state. Chomsky used the mathematical tools at his disposal, which meant rules and deductions. His framework was formal language theory. And the result was a theory of grammar restricted to syntactic structure, where meanig was at most an added-on projection rule. However, both syntax and semantics was part of the hidden revolution, but that was a fact that only became recognized through the work of R. Montague, a student of A. Tarski, in the 1960s. Tarski, who was widely recognized for his work on the semantics of higher order logic in the 1930s, was a colleague of Ajdukiewicz in Warsaw. They could have, but did not see the connection between categorial grammar and higher order logic. We had to wait for Curry, who was the first to do the science, and Montague, who twenty years later became the public face of the hidden revolution. This story and other parts of the relationship between mathematics and linguistics will be told at greater length in the chapter on language and logic.

Note. Parts of this chapter have previously been published in Fenstad (1996) and Fenstad (2004).

Grammar, geometry, and mind

Standard theory of grammar postulates the existence of two modules, one being a conceptual module which includes what is often referred to as knowledge of the world, one being a computational module which is concerned with the constraints on our actual organization of discrete units, such as morphemes and words, into phrases. Much of current theory is a theory of the syntax/semantics interface, i.e. a theory of how to connect grammatical space (the computational module) with semantical space (the conceptual module). In addition there has always been much work on the structure of grammatical space. However, remarkably little work has been devoted to the structure of semantical space. Even the Montague grammarians rarely make any use of the structure of their models; it is almost always possible to stay at the level of lambda-terms.

But this is a situation that is about to change. Many years ago I argued for the necessity of adding geometric structure to the models of formal semantics, see Fenstad (1978). I was guided by the "model-theoretic" approach to the methodology of science developed by P. Suppes, see the recent survey in Suppes (2002). At that time we had in linguistics two extremes: on the one side Chomsky and syntax, on the other side Thom and geometry; there was no mutual contact or any attempt to reach a common understanding. Today we have seen the development of a geometrical theory of meaning as an extension of formal semantics, see Gärdenfors (2000). Formal semantics may be adequate for current language technology, where the equation "model = data base" is a valid starting point, but it does not suffice for linguistics as part of cognitive science.

This chapter, which is basically a revised version of Fenstad (1998), traces a line from grammar to mind, the link being geometric model theory. We end up with a working hypothesis, which is admittedly a bit speculative at our current stage of knowledge, that a phenomenological theory of "mind" is nothing but a branch of geometric model theory. The geometry is the basic object introducing the lexical items, the rest – grammar, logic and mathematics – are formal tools used in the study of its structure.

Grammar and brain

We have in chapters 1 and 2 told a story top-down. But there can be no understanding of language and meaning without an insight into how the mind and brain are linked. The neuroscientist has now command of a vast and extremely detailed knowledge of brain structure. But when the linguist try to reach deeper into the mind and brain and the neuroscientist try to explain how higher cognitive functions emerge out of data and bare structure, we are in a somewhat uncertain middle ground. Symbols carry meaning, and a crucial part of the challenge of the middle ground is to understand how the interaction between symbol and meaning is rooted in the behavior of large and complex assemblies of nerve cells in the brain. Modern developments in logic have added to the syntax of grammar a (formal) meaning component. But this semantic structure, i.e. the model theory of logic, is a mathematical construction, and we need to understand how this structure can be used to explain the interaction of sound and meaning in the actual brain. This challenge has lead to an enrichment of logical model theory by adding geometric structure, this was our theme in chapter 2; the problem now is to understand how this geometry emerges from actual brain dynamics.

This is the structural view of models seen as "semantic space", a point of view that is essential to explain how common nouns, properties, natural kinds, attractors and brain dynamics are linked together. But logic has also a dynamical aspect necessary to analyze temporal reasoning and discourse. This use of logic must also be part of the attempt to understand language and meaning and how they are grounded in brain structure. In the last section, on adding time, we will give some references to this line of research.

A final word

The aim of this book has been to explore the relationhip between language, meaning and brain. I have tried to trace a line moving from grammar and logic, through meaning and mind, toward an attempt to link language and brain. Is there one connected story to tell? Strong reductionism would dictate a yes. But real phenomena are more complex than what simple scientific ideology would allow. In the "middle ground" between grammatical rules and brain anatomy – in the domain of ignorance – there are many levels and much to understand and explain.

1

Grammar

In most cases there is a reasonable correspondence between a subject matter and a science. Take physics as an example; there is little or no uncertainty about what the subject is and who are to be counted among the professionals. At every university there is always a central department of physics; nationally and internationally there are well defined professional associations, up to the level of the European Physical Society and the International Union of Pure and Applied Physics. There may be a bit of uncertainty about the exact line of demarcation between physics and some of the neighboring disciplines, such as chemistry among the natural sciences and electrical engineering among the technical sciences, but a combination of subject matter, cultural background and current sociology of the profession uniquely determines where you belong, whether you — to take one example with unexpectedly strong feelings involved — are a mathematical physicist or a plain mathematician working on mathematical problems in physics; no one would mistake the great Göttingen mathematician David Hilbert for a physicist even if he independently and at the same time as Albert Einstein arrived at the correct equations for the general theory of relativity.

The study of history offers another example. There is a core of the field focusing on political history, surrounded by spin-off specialities such as history of technology and even the history of science. Here also a combination of subject matter, cultural background and educational and professional behavior determines where you belong. One would never mistake a political scientist for a historian. Of course, in every-day life you may see a mathematician and a physicist, or a historian and a political scientists work peacefully side by side. But they know who they are and where home is; if challenged, they would at most describe their current activity as some 'problem-driven multidisciplinary project', if they are at all sensitive to what is politically correct. We may safely assert that there are few forces which are as strong as the gravitational pull of professional pride and property.

There may, however, be a counterexample. I have never been able to decide which science has the birthright to the study of human language and speech. Most professors of linguistics would claim this right for their own speciality; indeed, in most universities we will today find a department of linguistics — and is not linguistics the science of language and speech? Going beyond such superficial rhetoric we recognise that the case of language and science is remarkably more complex. In the European university tradition of the 19th century there were no separate departments of general linguistics. Language and literature were always the organizational unit, and one such unit existed for each language or language family. It was the needs of the *gymnasium* which determined the structure of university departments in the humanities. In the emerging national states of Europe there was a special emphasis on the national language and literature and this had implications for education on all levels. The classics did not disappear; there were separate departments for the classics as well as for the major modern foreign languages. For the rest there were various *ad hoc* arrangements; e. g. in the Nordic countries there would typically be a single unit for all the slavonic languages. The subject matter was dominated by the humanistic heritage of philology, with the usual emphasis on textual interpretation. After the "discovery" of sanscrit the linguistics curriculum was enriched by historical and comparative studies. Such was — in simplified terms — the 19th century picture; for a comprehensive survey of the period I recommend H. Pedersen (1972) *Linguistic Science in the Nineteenth Century*. But language, in particular seen as a cultural phenomenon, soon came into focus in other disciplines. There is an obvious link to anthropology; indeed, general linguistics, as distinct from philology and comparative studies, sometimes found its first academic home in schools or departments of anthropology; this was not uncommon in the anglo-american tradition. The reason is perfectly clear, language as a general cultural phenomenon cannot be bound by a tradition with roots in the study of Latin and Greek.

With the emerging social sciences we also saw other interests taking claims on language as a field of study; psycho-linguistics focused on language and the individual; socio-linguistics on language and society. It always was, and still is, unclear where the new hyphenated disciplines should belong. One should be careful in making bold generalizations, but my impression is that language as a cognitive phenomenon was more central to psychology than language as a social phenomenon to sociology.

This is not the end of the story. The study of speech was always an integral part of linguistic. But with the advent of 'modern' telecommunication, i. e. radio and telephone, the transmission of speech, seen as an important example of the technology of signal transmisson, has become a recognized engineering speciality. Originally there were few links between the technology and the 'core' science of linguistics. Today telematics, seen as part of the expanding information technologies, has progressed beyond the original signal transmission stage and faces new demands for more comprehensive technologies for language processing. One would assume that such technologies should rest on a broad science of language, and there is currently much activities devoted to a proof of this point. The matter, however, needs careful thought; I shall return to the issue at the end of this chapter. At the moment I will only note that speech recognition, which is a central component of the technology, is almost totally independent of current theoretical and computational linguistics. We do have effective systems for speech recognition, but that technology is basically statistical in nature, making use of Bayesian transformations, hidden Markov-chains, frequency analyses of large text corpora — but it makes almost no use of grammar. We should, however, note that this is a situation that may change. Current developments of neural networks for format of phonemes may be important for future technology; see the section below on language technology.

We have placed the study of language at the cross-road of various humanistic, social science and engineering disciplines. If we also want to see language as a brain-determined activity, we need to add several of the natural and medical sciences. It is for many reasons important to understand how the humanities and the sciences are interacting in modern society, but rather than to reflect in general on this, I have chosen to concentrate on the specific case of language. This adds the richness of a specific context; I would also argue that whatever general points can be made on the relationship between the humanities and the sciences, can be made with equal force in our special case.

In the catalogue of contenders for the ownership of the science of language the reader may have noticed the omission of one discipline — logic. Language and logic has been inseparably intertwined in the European intellectual tradition. It has not always been an easy relationship, but there were always issues of substance whether they were friends or foes; some observations on the history of this relationship will be covered in the next section.

1.1 Some observations on the history of grammar and logic

The starting point for this reading of the history is, of course, the Greek heritage, although the history of the systematic study of language is older and has more complex roots than the Greek experience. The systematic study of language was part of the general theoretical awakening in ancient Greece, with Plato and Aristotle as the central actors. But whereas the history of mathematics and the natural sciences always has been an integral part of our cultural history, the history of linguistics has remained a concern for the specialists. This is a bit curious seen on the background of the dominating role of grammar, logic and rhetoric as part of the *trivium* of classical learning, the basis for higher education in Europe through the centuries.

After Aristotle there was a separation of ways in the study of grammar. One line of development was the philosophical or theoretical study developed by the Stoics and transmitted to the medieval grammarians through the rediscovery of Aristotle by Boethius. The other line was the Alexandrian tradition represented by grammarians such as Dionysius Thrax and Apollonius Dyscolus. This was the so-called literary or descriptive way which was particularily aimed towards the teaching of language and which built upon extensive examples drawn from the classical literary heritage. We can already at this point see a first example of a split in linguistics between a data-driven and a theoretical approach.

This account is not intended as a general history, but represents a personal view of the field. I shall therefore focus on what is necessary for the later parts of the story. My next stopping point is at the beginning of the fourteenth century. Grammar had in the early medieval period been dominated by the descriptive line, focusing in particular on language teaching. Time was now ripe for a rebirth of a theoretical grammar. We can not follow all the complexities of the historical development; the interested reader could consult the edition of *Grammatica Speculativa of Thomas of Erfurt* by G. L. Bursill-Hall (1972). For our purpose I note that a group of grammarians, the so-called *modistae*, with Thomas of Erfurt as a leading representative, managed to give a theoretical foundation to an existing descriptive practice. In doing so they buildt upon the rediscovered Aristotle, but went far beyond in creating a new theory. With the help of generous hind-sight we can recognize in the theoretical analysis of the *modistae* both a grammatical surface structure and a semantical-logical deep structure. The grammar of the modistae is beyond doubts a high point in the scholastic edifice of learning. Today their theory is for many a closed world, imbedded as it is in scholastic philosophy. But in my judgement Thomas of Erfurt is exemplary in his attempt to give a connected account of structure and meaning.

For the current observer of the linguistic scene it may look strange to stop at the strong medieval brew of Thomas of Erfurt. Why not, following current fashion,trace our 'revolution' in linguistics and language engineering to the clear cartesian foundation found in the Port-Royal grammar of ca 1660? I would have liked to agree; I have a high opinion of the *Grammaire Général and Raisonné* by the Port-Royal jesuits Lancelot and Arnauld. But one need to show some care with the historical scholarship of the much celebrated book *Cartesian Linguistics* by N. Chomsky (1966); see e. g. the critique in G. A. Padley (1976) *Grammatical Theory in Western Europe 1500-1700*. We may now with some confidence claim that it was the *modistae* and not the Port-Royal grammarians, who had the fresh and original insights, and that it was they who had a perspective which again today is productive in theoretical linguistics. But before this could be appreciated linguistics once more divorced itself from logic and turned back onto a descriptive road.

We next make a jump to the 1930s. The excessive historical and comparative studies of the 19th century had by that time been replaced by a structuralist approach, which was initiated by the great Swiss linguist Ferdinand Saussure. We shall take the well-known American linguist Leonard Bloomfield as our witness. As a structural linguist he occupied a central position in his discipline toward the end of the 1930s. Philosophically he subscribed both to empericism and behavioralism. In an article *Linguistic Aspects of Science*, Bloomfield (1955), which he contributed to the *International Encyclopedia of Unified Science* he gave an almost militant expression to his convictions. In his brief survey of the history of the subject he also starts with the Greeks, but his tone is severely critical, accusing them of serious mistakes due to their unscientific (read: metaphysical) definitions. Following Bloomfield modern linguistics should trace its roots to the rediscovery of sanscrit and to the ancient grammar of that language by the Indian grammarian Panini. We see in Panini, still according to Bloomfield, a descriptive — even 'scientific' — grammar, rid of philosophical assumptions. This kind of grammar set an example for the description of native Indian languages in America, and it provided a historical foundation for the data-driven and inductive approach of the structuralist school in American linguistics. Bloomfield's contribution to the *Encyclopedia* is fascinating reading. If we only could believe in him, the final victory was at hand. But read on the background of current insight into the cognitive sciences, his behavioristic foundation for linguistics is seen to be a remarkably primitive stimulus-response construction.

American structural linguistics saw itself as the culmination of linguistics as a science. The book that was intended to put the last stone in its proper place,was *Methods in Structural Linguistics* by Zellig Harris (1951), which appeared in 1951. The methodology is sketched out in an introductory chapter. The procedure is strictly inductive and based on a comprehensive distributional occurrence analysis. The structure of the particular language under study must be read off from the inductive analysis. The theory of linguistics has *de facto* been reduced to a procedure for language description. What remains of meaning is at most an epiphenomenon, a pure case of stimulus-response behavior, which is objectively observable by the field linguist. There is no universal theory of grammar and meaning. We should, however, add that this somewhat polemical point does not at all deny the importance of corpus studies for current linguistic theory and technology.

The brevity of this description is unfair to what Harris did. He was a great linguist with fundamental insights to his credit; see the review article by J. Goldsmith in *Language* vol 81 (2005). Experimental design and descriptive procedures are an important part of any empirical science; read in this light his contributions are important. He also indicated the need for a mathematical study of the elements which emerged from the inductive analysis, and he pointed to the importance of mathematical transformations in the final structural description (e. g. an active-passive transformation). The seeds of a generative analysis is thus implicit in some of his theoretical considerations.

We now move forward to the year 1957. This year marks the beginning of Noam Chomsky's revolution of linguistic science. We should, however, note that there were some exceptions. The revolution was in the beginning an American phenomenon, there were European traditions which long remained outside the revolutionary circle. But the impact of Chomsky fully justifies the word 'revolution'.

The ground was well prepared. The philosophical foundation, i. e. the extreme behaviorism of Bloomfield, had started to crumble. Skepticism concerning the possibility of a pure inductive determination of structure was also more pointedly raised; some critics accused the structuralist for in secret to look at meaning and other mental phenomena in the determination of structure; they pointed especially to the supraseg-mental analysis of intonation. On the other hand, one had seen the start of a formal analysis already in the work of Harris, and modern develop-ments in mathematical logic, in particular, in recursion theory and the theory of formal languages, provided the tools necessary to translate such linguistic insights into "hard" science. The bits were assembled into a coherent picture by Noam Chomsky. First in a short monograph *Syntactic Structures* from 1957; a few years later in a comprehensive and convincing manner in the book *Aspects of the Theory of Syntax*, Chomsky (1957, 1965). This is not the place to give an introduction to Chomskian generative linguistics. We only note that within a couple of years there had occurred a revolution in linguistics. A historian of science would probably see some parallels between Chomsky's Aspects and the influence it had on the linguistics of its time and Thomas of Erfurt's *Grammatica Speculativa* and the influence of the modistae.

The analogy is only partial. Chomsky argues in principle for an equal status of a syntactical, a phonological and a semantical component, but in his actual model-building it is the syntax and the phonology which dominate. The reason may have to do with the mathematics available at the time; this is a point I shall return to later.

Here I wish to emphasis that the Chomskian revolution was not only an internal issue of content, it was equally a cultural and socio-logical revolution; the reader may wish to consult *Linguistic Theory in America*, F. J. Newmeyer (1980). But simultaneously with the Chom-skian transformation of the linguistic landscape there occurred another 'revolution'. It was hardly visible, and it took the linguistic profession almost twenty years to digest what had happened. In retrospect the sec-ond revolution can make a stronger claim of being the 'true' successor to the revolution of the *modistae* of medieval times.

Modern logic is more than axioms and deductions. Semantics was always an integral part; in modern times we can draw a line from the German logician G. Frege via the Polish logician A. Tarski to the cur-rent theory. John Stuart Mill wrote in 1867 that 'the structure of every sentence is a lesson in logic', but the structure of logic seemed at first to be much too rigid to serve as a tool for linguistic analysis, except at a sufficiently deep logical level. The logical empericisist of the 1930s also held natural language in low regards, as the following programmatic pronouncement of C. W. Morris, one of the chief architects behind the *Encyclopedia*, bears witness to: '... it has become clear to many persons today that man — including scientific man — must free himself from the web of words which he has spun and that language — including scientific language — is greatly in need of purification, simplification and systematization. The theory of signs is a useful instrument for such debabelization', Morris (1955). Let us add that logical syntax and semantics in the style of Rudolf Carnap (1937) was to be the main tool in this task; everything that did not fit into this scheme should be demoted to a behavioristic stimulus-responsemodel. It is remarkable that Bloomfield was a co-contributor to the same unified programme.

Tarski denied that his logical analysis in *Wahrheitsbegriff* had any relevance for the understanding of truth and meaning in natural languages. This may be true. Tarski's analysis in *Wahrheitsbegriff* has remained a cornerstone in formal semantics; for a recent up-date on truth and partiality, see the author's article *Partiality* in the *Handbook of Logic and Linguistics* (van Benthem and ter Meulen (1997)). But if we are concerned with the broader interaction between logic and linguistics, we see a missed opportunity. What had happened if Tarski had paid attention to what his Warsaw colleague K. Ajdukiewicz tried to do in his article *Die Syntaktische Konnexität* from 1935? Tarski and Ajdukiewicz belonged to the same group of logicians in Poland in the mid 1930s. In the two papers mentioned above there are mutual references. Tarski has several references to Ajdukiewicz, also inside the section where he discusses "colloquial" languages (see footnote 1 on page 161 in Tarski (1956)), but not to *Die Syntaktische Konnexität*. Ajdukiewicz has a reference to the *Wahrheitsbegriff* paper (see footnote 1 on page 209 of Ajdukiewicz (1967)) where he notes the similarity between his theory of semantic categories and the hierarchy of logical types. Indeed, there are similarities. Both systems derive from the Russell-Whitehead theory of types; directly in Tarski's *Wahrheitsbegriff* and indirectly via Lesniewski (1929) in Ajdukiewicz's *Syntaktische Konnexität*. The link is – in hindsight – all too obvious. The categorial grammar of Ajdukiewicz is an applicative system corresponding to higher order logic. It was precisely for these logic systems that Tarski developed his concept of truth. Contrary to both Tarski and Bloomfield (and their fellow travellers Morris and Carnap) it would have been possible in the 1930s to build a viable link between logic and linguistics. We could have had a modern version of the great synthesis of the medieval *modistae*; see the *Grammatica Speculativa* by Thomas of Erfurt from 1315. But both the methodological dogmas of the Vienna Circle and the strict inductive methodology of American structural linguistics forbade such speculations.

But mathematical model-building is more than the naive formalization of the early empiricists. One of the first persons to apply the whole range of tools of modern logic to the study of natural languages was the German philosopher of science Hans Reichenbach, who included a chapter on 'conversational languages' in his textbook *Elements of Symbolic Logic* from 1948. The particular tool that Reichenbach used was a version of higher order logic extended by certain 'pragmatic' operators.

From about the same time we find a contribution by Haskell B. Curry, an American logician and one of the last to receive his doctorate in Göttingen under the supervision of Hilbert. Curry lectured on his results in 1948, but a paper was published only in 1961, Curry (1961). In this work Curry presents an analysis of traditional grammar using combinatory logic. The analysis is related to, but independent of, the 'categorial analysis' presented in the 1930s by the Polish logicians S. Lesniewski and K. Ajdukiewicz. Curry's main advance on the Polish is that his combinatorial logic has a naturally associated semantics in terms of a higher order functional structure.

This was in retrospect the second 'revolution'. But as remarked above it was hardly visible. Curry remained unpublished; Reichenbach's book appeared a few years before Harris' *Methods in Structural Linguistics*, and must for the contemporary linguists have represented a step back to the *a priorism* of logic and metaphysics. But linguists ought to have read their Reichenbach with more attention. They would then have noticed that he gave a first satisfactory analysis of meaning of natural language, using higher order logic. He broke through the narrow framework of standard predicate logic, which long had been recognized as inadequate for linguistic analysis The pragmatic operators he introduced, were a first approximation to later developments in intensional logic. The analysis of time which he gave, is still the foundation for present day linguistic theories. The main complaint against Reichenbach, but not to the same degree against Curry, is the lack of an explicit syntactic component. But even here the tools were available at the time; I am thinking of the categorial grammar of the Polish logicians.

Few people have read Lesniewskis paper from 1929 on *Grudzüge eines neuen Systems des Grundlagen der Mathematik*, his ideas were made known through the work of Ajdukiewicz *Die Syntaktische Konnexität* from 1936. This line of development was carried forward by the Israeli philosopher of science Y. Bar-Hillel, who wanted to use categorial grammar in machine translation around 1950; more on this below.

The second revolution remained at the time a possibility. The necessary elements were at hand, but no one saw how to fit the pieces together in a picture that would convince the linguistic community. It could have been Bar-Hillel, the logician who knew his Reichenbach and who wanted to apply categorial grammar to machine translation. But Bar-Hillel, after the early failure of machine translation, became toward the end of the 1950s an enthusiastic supporter of Chomsky; thus categorial grammar, combinatory logic and higher order structures disappeared from center stage. Posterity may regret what happened. We may feel that the lack of insight was remarkable, in particular, since all pieces had the same ancestry, which was the theory of types of the *Principia Mathematica* of Russell and Whitehead from 1910. It took twenty years before the pieces were finally put together. This was achieved in a series of works by Richard Montague from the mid 1960s; see his collected essays *Formal Philosophy*, Thomason (1974). Why did some fail and other succeed? This is an issue which involves both mathematics and the sociology of the profession.

1.2 Grammar and the mathematical sciences

We shall look in some details at a few recent episodes in the sometimes turbulent relationship between linguistics and the mathematical sciences. For present purposes our starting point will be the *de facto* seperation of ways of the mid 1930s. In linguistics American structuralism was a leading paradigm. Linguistics was reconstructed as an empirical science, proceeding in an inductive way in its task of extracting structure from data. Meaning was an added epiphenomena of objectively observable stimulus-response behavior. The other partner of our story, philosophy, always has been more confused, but one leading paradigm of the 1930s was the school of logical empiricism. A leading spokesman, first in Europe, later in the US, was Rudolf Carnap, who also was one of the main forces behind the *International Encyclopedia of Unified Science*. Carnap obviously would recognize the need for an empirical study of language as a cultural phenomenon, but he would no doubts subscribe to the pronouncement of Morris that 'language — is greatly in need of purification, simplification and systematization'. Carnap's own contribution to the needed 'debabelization' was the severely formalistic tome on *The Logical Syntax of Language*. Hans Reichenbach was also one of the leading scientific philosophers, well-known for his contributions to the foundation of probability, to the study of space-time and the foundation of quantum mechanics. Highly respected is also his study on the irreversibility of time. We have praised his contribution to linguistics, but to his contemporaries the chapter on 'conversational' languages was only a minor, perhaps, even curious episode in a rich scientific career.

Not every linguist marched under the banner of structuralism. Let me note one exception, the Danish linguist Otto Jespersen. He wrote in the 1920s a book *The Philosophy of Grammar*, Jespersen (1975), which was at odds with the structural thesis and much more in line with traditional grammar. The book has been reprinted a number of times and is still read today. In 1938 he published a short volume on *Analytical Syntax*, Jespersen (1969), in which he tried to give a formal expression to his views. But that book — even if it has been recently reprinted, since Jespersen is a famous linguist — is curiously barren in content. My explanation is simple; Jespersen did not know the kind of mathematics necessary for his task. He belonged to a long and distinguished tradition of Danish linguistics, which was dominated by the historical and comparative themes of the 19th century, but which gradually changed to the structuralist point of view. Jespersen had his independence of mind, but he was bound by his educational heritage. In no way was the acquisition of a broad mathematical culture part of the upbringing of a European linguist.

It is instructive to contrast this episode with the case of Haskell B. Curry. We met Curry above as one of the silent heroes of the second revolution. His starting point was his familiarity with traditional grammar, and he possessed the mathematical tools necessary to give a formal expression to his insight into grammatical structure. To be slightly technical, traditional grammar is cast in a subject-predicate form, in fact, the elementary grammatical operations can be seen as examples of a function-application formalism. Combinatory logic, which was the speciality of Curry, is an applicative system particularily well suited for the formalization of traditional grammatical operations. This is a feature which it shares with the categorial grammar of Ajdukiewicz. But combinatory logic adds more. It has a natural associated meaning theory in the form of a higher order functional structure. Thus Curry was able to provide a combined syntactical-semantical analysis of a fragment of English. Curry was no linguist, but his familiarity with traditional grammar was sufficient to produce a significant scientific insight. Curry had little impact on the linguistic community; he was quite isolated in his search for linguistic structure, neither linguists nor logicians paid much attention.

The state of Pennsylvania in the US was Curry's home territory. A few years later the same geographic location was the scene of the 'visible' revolution in linguistics. Noam Chomsky was in a somewhat unusual sense the student of Z. Harris at the University of Pennsylvania; see the introductory interview in the collection *A Chomsky Reader*,

Peck (1987). I do not know the exact influence of Harris on Chomsky, but as we noted above there are mathematical seeds in Harris' *Methods in Structural Linguistics* which in a full-grown way are present in Chomsky's *Syntactic Structures*. Similarily, I do not know if there were any serious contacts between Curry and Harris. Curry's paper was presented in a seminar for linguists. If it had provoked any response, Harris would have known. I find it fascinating to see the celebrated 'structural linguist' Harris interpolated in this curious way between the 'logician' Curry and the 'generative linguist' Chomsky in the late 1940s — early 1950s in the state of Pennsylvania.

Chomsky succeeded, partly because he was a master of the mathematics he used; indeed he made himself several important contributions to formal language theory. But part of the success was also sociological. The linguistic community was ripe for the elevation of their field of study to one of the 'exact' sciences; see the account in Neymeyer's book. Let me, however, add one slightly critical observation on this success story. Chomsky was, as one always is, constrained by the mathematics he knew. Meaning was always a recognized part of the Chomskian paradigm, but the mathematics he mastered, forced an almost exclusive attention to syntactical matters. Remembering Chomsky's attempt to link his linguistics theories to the Port-Royal tradition, one conjectures that he would have liked to se a full-blown mathematical development of meaning. But Tarskian semantics was still in its infancy around 1950, even Reichenbach had to resort to some *ad hoc* trickery to give an account of meaning.

Richard Montague was a student of Tarski in Berkeley in the late 1950s. He worked independently, but seen from our vantage point his contribution consisted mainly in making a coherent pattern out of the elements found in Ajdukiewicz, Curry and Reichenbach. I particularily recommend his paper on the *Proper Treatment of Quantifiers in English*, Montague (1974), as a substantial sample of what he did. Montague succeeded where Curry failed. We can ask why; the insight and power of their respective mathematical analyses are comparable; they both could have succeeded. But Montague had an interpreter whereas Curry stood alone. Barbara Hall Partee is a leading American linguist. She was professor in the Linguistic Department of the University of California at Los Angeles at the same time as Montague was a professor of philosophy there. Whereas Montague saw his work as an exercise in formal philosophy, Partee saw the linguistic significance of his work. Through a series of studies she made his work accessible to the linguistic community. She also made important new contributions, highlighting the linguistic relevance of the formal apparatus. With her stamp of approval linguistics had to take notice.

There is one more episode to tell; this time a European, and it involves both a linguist and a mathematician. The linguist is L. Tesniér, a French scholar working outside the main intellectual centers. Whereas the structure of Chomskian syntax was rather Aristotelian in form, Tesniér attempted to find a pure relational form to the syntax, putting the verb phrase into a central position; see his book *Eléments de Syntaxe Structural* from 1958; in this respect he is in the line of Curry, Reichenbach and Montague. The syntactical studies of Tesniér has inspired other European linguists, particularily in Germany; we mention the *Dependenztheorie* as one important example. Tesniérs work also served as inspiration for some interesting geometric reflections on grammar by the French mathematician René Thom. In two papers from around 1970, *Topologie et Linguistique* and *Langage et Catastrophes: Elements pour une Semantique Topologique*, he used the syntactic analysis developed by Tesniér as his linguistic background, Thom (1970, 1973). The central role of the verb phrase in Tesniér is in Thom translated into a classification of singularities in a suitably derived energy surface. This may sound odd, but it is a 'correct' step toward a comprehensive theory of meaning in natural languages. Thom's work did not have the influence on theoretical linguistics that it merits. There seems to have been an incompatability of minds. Thom severely criticised the generative approach as being totally inadequate as a theory of linguistic meaning. Linguists, unfamiliar with the mathematics, saw little relationship between Thom's 'speculations' and their science. Do we once more see a weakness of the European university tradition, which has been dominated by a strong sense of professional purity,which in turn has tended to encourage a certain 'inwardness' in the education of young scholars. It seems that the American Graduate School at its best was remarkably effective in creating a broader intellectual community. This is not a question of fashionable multi-disciplinarity, but of the plain task of learning the tools of many trades. Barbara Hall Partee had a sound training in mathematics in the graduate school of MIT.

1.3 Remarks on the formal theory of grammar

It is time to fill in some technical details. The theory of grammar was never a unified science. There seems, however, to be one shared assumption. Most theoretical linguists will subscribe to the following point of view expressed by the Dutch linguist Jan Koster in a lecture some years ago; Koster (1989):

> ... we have to make a distinction between a computational structure
> and a more or less independent conceptual structure ... The concep-
> tual module includes what is often referred to as knowledge of the
> world, common sense and beyond. It also includes logical knowledge
> and knowledge of predicate-argument structure. The computational
> module concerns the constraints on our actual organization of discrete
> units, like morphemes and words, into phrases and constraints on re-
> lation between phrases (Koster (1989), p. 593).

Later he adds (Koster (1989), p. 598) — and this is not a view universally accepted by linguists:

> If the two basic modules are autonomous and radically different in
> architecture and evolutionary origin, there must be ... an interface ...
> My hypothesis is that the interface we are looking for is the lexicon ...

.

A major part of post-Chomsky linguistic theory has been devoted
to the investigation of the computational module. Post Montague we
have seen an increased interest in the conceptual module, or semantic
space. In a certain sense the computational module of the linguist corresponds to syntax as used by the logician and the conceptual module
corresponds to the use of semantics or model concept. This is a rough
indication and the usage may differ between different authors.

Both in Curry (1961) and Montague (1974) we have two major parts
to the model: One is an *extended categorial grammar*, corresponding to
the computational module of Koster, which gives a syntactic analysis of
a well-defined fragment of a natural language. The second part, which
corresponds to the conceptual module of Koster, consists of a *translation* into a system of higher order (intensional) logic. The translation is
required to observe the principle of *compositionality*, which means that
it is (in an appropriate sense) a *homomorphism* between the syntactic
"algebra" and the semantic "algebra".

In his analysis Montague implicitly used the notion of *generalized
quantifier* to give a uniform treatment of *noun phrase*; see Fenstad
(1978) and Barwise and Cooper (1981). Thus if *john* and *mary* are
items of the lexicon, they will not be translated as constants j and m
of individual type, but be translated as:

john	translates to	$\lambda P.P(j)$
mary	translates to	$\lambda Q.Q(m)$

The term $\lambda P.P(j)$ corresponds in the semantic interpretation to the
set $\{X : j' \in X\}$ in the model, where j' is the value of the constant j.
Note that for simplicity we use an extensional version of higher order
logic.

The sentence *john loves mary* has in the Montague model the following syntactic analysis:

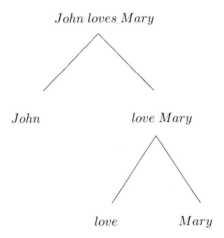

Compositionality, or the homomorphism criterion, demands a translation of *love mary* as an application of the translation of *love* to the translation of *mary*,

$$love\ mary \qquad \text{translates to} \qquad love\,'(\lambda Q.Q(m)),$$

where *love'* is an appropriate constant of the logic as translation of the lexical item *love* of the natural language fragment. In the next step compositionality dictates the following translation:

$$john\ loves\ mary \qquad \text{translates to} \qquad (\lambda P.P(j))(love\,'(\lambda Q.Q(m))).$$

Compositionality requires that every node in the syntactic tree has an associated object in the logic and the semantic algebra. And the syntactic analysis tells us which type to assign to the associated object. Thus the primary translation *love'* is no longer a relation between individuals, but a more abstract function. Not everyone in love would be happy with this analysis!

There is an answer that some people has taken to be a success of the Montague analysis; see Fenstad (1978). The logic may have a basic constant *love** denoting a relation between individuals. Using lambda-abstraction we can then define a new constant *love'* of the appropriate type by

$$love\,'(\mathbf{P})(x) \qquad \text{if and only if} \qquad \mathbf{P}(\lambda y.love\,^*(x,y))$$

This "lifted" *love'* has the correct type to match the syntactic analysis:

$$love'(\lambda Q.Q(m))(j) \quad \text{conv. to} \quad (\lambda Q.Q(m))(\lambda y.love^*(j,y))$$
$$\text{conv. to} \quad (\lambda y.love^*(j,y))(m)$$
$$\text{conv. to} \quad love^*(j,m)$$

This is what Tarski and Ajdukiewicz could have done in the mid 1930s. Everything was in Curry (1961), a paper which was written in the late 1940s, but largely neglected. The work of Curry did not receive the attention it merited at the time, and it was the later contributions of R. Montague that revitalized the links between logic and linguistics; see the reprint collection Thomason (1974). This represents, in fact, another piece of "deconstruction": The tree structure of categorial grammar is mapped into the relational form of model theory. One should not be confused by the complexities of the lambda-terms of logic. Higher order logic defines the map between the tree structure of categorial analysis (the subject-predicate form) and the flat relational form of model theory – the higher order logic is the tool, not the substance. This use of logic as a connecting map has led to a number of interesting insights. Particularly important is the structural identification between noun phrases and generalized quantifiers; see Fenstad (1978) and Barwise and Cooper (1981). The link between generalized quantifiers and noun phrases attracted much interest and became a central part of the connection between logic and grammar. But we should appreciate that quantifiers in logic and linguistics is a much broader theme; see the recent book by S. Peters and D. Westerståhl (2006).

Returning briefly to higher order intensional logic, we recognize that it is a powerful and elegant mathematical tool. But when this instrument is used to establish a map between syntax and meaning there is a tendency to formalize too much. This was a criticism voiced, among others, by J. Barwise and R. Cooper in their study of the relationship between noun phrases and generalized quatifiers; see Barwise and Cooper (1981). Further reflection on the proper semantic structure for the study of natural languages led to the subsequent development of situation semantics; see J. Barwise and J. Perry (1983). Language and formal semantics have been an area of great activity. The full story remains to be told, see however the *Handbook of Logic and Language* (edited by J. van Benthem and A. ter Meulen, 1997) with many useful reviews of Montague grammar, categorical grammar, types, partiality, generalized quantifiers and related topics. Some of the 'classics' are collected in *Formal Semantics: The Essential Readings* (P. Portner and B. H. Partee, 2002). A textbook exposition can be found in Carpenter

(1997). And a broad historical review of the interaction between logic and language can be found in the survey paper *Logic and Linguistics in the Twentieth Century* (A. Lenci and G. Sandu, 2005).

The notion of *the connecting sign* was a central element of Koster's analysis. And, indeed, the nature of the *syntax-semantics interface* has been a major issue in recent linguistic theory. In early Chomsky (see his *Aspects of the Theory of Syntax* from 1965) the connecting sign between grammatical deep structure and semantical representation was basically an arrow decorated with a name (a so-called projection rule) – but with no particular content; we noted above that formal semantics was not well developed at the time when Chomsky started his work. With Montague, a student of Tarski, semantics is part of the tool-box, and, as we have seen above, in his model the connecting sign is most readily identified as the translating formula in higher order intensional logic.

A major alternative to formulas as connecting signs is the use of attribute-value structures. This is an approach which in a very concrete sense uses a lexical sign as the connecting element between the computational and conceptual modules. An early and important example of such a theory is the *Lexical-functional grammar* (LFG) developed by R.Kaplan and J.Bresnan (1982). We shall examine the situation in more details.

The *Center for the Study of Language and Information* (CSLI) was founded at Stanford in 1983. The opening year of the Center was marked by a general wish to explore the connection between the many disciplines present. LFG was the prominent syntactic theory at CSLI. Situation semantics had a similar status on the meaning side. Syntax and semantics must be related. Thus the question of how to interpret the functional structures of LFG in situation semantics became urgent. The solution, simple when first recognized, was the concept of situation schemata; see Fenstad *et al* (1985, 1987). This is a representational form derived from the f-structures of LFG. And in contrast to the lambda-terms of Montague grammar, questions of efficient computability was always an important concern. The technology of situation schemata was later adopted by Pollard and Sag (1987) in their development of HPSG. Today attribute-value formalisms have emerged as one of the most important frameworks for grammatical analysis. This is by now text-book material; for LFG see J. Bresnan, *Lexical-Functional Syntax* (Bresnan, 2001); for HPSG see *Syntactic Theory: A Formal Introduction* (I. A. Sag and T. Wasow, 1999); see also the discourse representation theory of Kamp and Reyle (1993). It is also of interest to note that some current developments in the Chomskian tradition represents a move toward an attribute-value formalism; for further details see R. Jackendoff (2002) and P. W. Cullicover and R. Jackendoff (2005). There is also an extensive literature on the formal aspects of attribute-value systems, see M. Kay (1979, 1992), J. E. Fenstad, B. Vestre and T. Langholm (1992), the *Handbook* survey by W. C. Rounds (1997), and the books by B. Carpenter (1992) and B. Keller (1993).

Every theory of language has as its ultimate goal an account of the link between linguistic structure and meaning. But different theories differ as to where to locate the connecting sign along the syntax-semantics axsis. Chomsky's *Aspects* is biased toward the syntax end. To a certain extent this is also true of classical LFG. HPGS is more balanced between the two components. From our point of view Montague's theory is above all a theory of the connecting sign. The grammar is simple and the theory makes almost no use of the model structure. Much early theory in the Montague tradition consisted in the manipulation of lambda-terms.

How does this compare to the pairing of LFG and situation semantics? We need to be a bit more specific about the semantics, see Barwise and Perry (1983). The starting point is a multi-sorted structure

$$M = (S, L, D, R),$$

where S is a set of *situations*, L is a set of *locations*, R is a set of *relations* and D is a set of *individuals*. Note that in situation semantics all basic types are primitive, which, in particular, means that a relation is not a set of n-tuples of individuals. Sets of tuples may be used to classify relations, but, as argued in Barwise and Perry (1983), this is not sufficient as analysis in a broader cognitive context. *Basic facts* are either positive or negative,

$$r, a_1, \ldots, a_n; 1$$
$$r, a_1, \ldots, a_n; 0$$

where $r \in R$ and $a_1, \ldots, a_n \in D$. *Partiality* is present in the format since we do not necessarily have either $r, a_1, \ldots, a_n; 1$ or $r, a_1, \ldots, a_n; 0$, for all n-tuples $a_1, \ldots, a_n$. Facts may be *located*,

$$\text{at } l: \ r, a_1, \ldots, a_n; \ i \quad (i = 1 \text{ or } 0)$$

where $l \in L$ is a connected region of space-time. A *situation* is determined by a set of located facts of the form

$$\text{in } s: \text{at } l: \ r, a_1, \ldots, a_n; \ i.$$

The main contribution of Fenstad et al. (1985,1987) was a formal construction of a method which to every sentence of (a fragment of) a natural language (taken e.g. from some text corpus) gave an interpretation of that sentence in a system of situation semantics *via* its associated situation schema; the technical details are spelled out in the remark below. From one point of view this is a piece of theoretical linguistics, but there were also some early applications of the techniques to natural language technology, in particular, to question-answering systems, see E. Vestre (1987) and Fenstad et al. (1992), and to machine translation systems, see H. Dyvik (1993).

Cognitive Grammar is an interesting exception to this tradition. Here we see a shift from the syntax to the semantics end of the connecting axsis; for a development of this theory see R. W. Langacker (1987, 1991). In the earlier forms of the theory we still recognize two components, one phonological and one semantical. There is also a connecting symbolic structure, which in some sense is a pictorial attribute-value structure. In a purer version of the theory the phonological module is subsumed under the semantic one: "... phonological space should instead be regarded as a subregion of semantic space (Langacker (1987), p.78)". Cognitive grammar is a minority view in linguistics, but it may have a valid point in its revolt against the combinatorial dominance within current linguistic theory; for a discussion of this point and the geometrization of thought see Gärdenfors (2000).

Contrary to cognitive grammar the standard view in linguistics has been that syntax is the only input to semantic interpretation. This is particularly true of categorial grammar and theories in the Montague tradition which are based on type theoretic formalisms. In these theories meaning is a homomorphic image of the syntax in a very precise sense: the interpretation function is a homomorphism (i.e. a structure preserving) map between the algebra of formulas and the set-valued algebra generated by the model as explained in the example above. The world is what can be expressed in your language. We can read a different philosophy behind the attribute-value formalism. This formalism can be looked upon as a constraint-based view of the relationship between linguistic form and meaning.

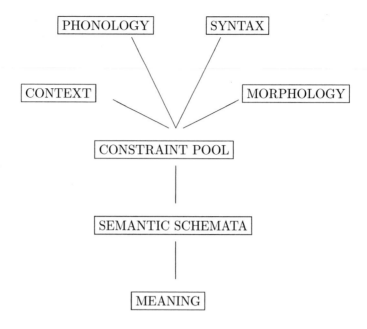

In this view all aspects of linguistic form, such as phonology, syntax, and morphology as well as context contribute to a combined set of constraints, which in turn determines the meaning of an utterance. In theories of the LFG variety the constraints are given in form of equations and the resulting linguistic sign or attribute-value matrix represents a consistent solution to the constraint equations; for an extended discussion see chapter II of *Situations, Language and Logic* (Fenstad et al. (1987)).

A technical remark. For the curious reader we include at this point some formal details taken from Fenstad (1988).

A *situation schema* is a complex attribute-value structure computable from the linguistic form of utterances and with a choice of attributes matching the primitives of situation semantics. The basic format is thus

$$\begin{bmatrix} REL & - \\ Arg.1 & - \\ \vdots & \vdots \\ ARG.n & - \\ LOC & - \\ POL & - \end{bmatrix}$$

Here the attributes REL, $ARG.1, \ldots, ARG.n$ and LOC correspond to the primitives of *relations*, *individuals* and *locations*. POL, abbreviating *polarity*, takes either the value 1 or 0. The values in the schema can either be atomic or themselves complex attribute-value stuctures. The value of LOC is always complex.

As an example let us write down the schema corresponding to the sentence *John married Jane*: Let Φ denote this sentence, then $SIT.\Phi$, the associated situation schema, will be:

SIT.Φ

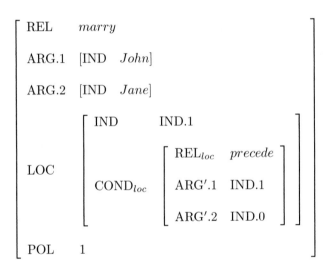

The situation schema is a representational form which can be interpreted in a system of situation semantics relative to an *utterance situation u* and a *described situation s*. The utterance situation decomposes into two parts

d *discourse situation*
c *the speaker's connection*

The discourse situation contains information about who the speaker is, who the addressee is, the sentence uttered, and the discourse location. The speaker's connection is used to determine the speaker's meaning of lexical items; this we shall shortly explain.

A map g defined on the set of indeterminants of the value of LOC and with values in the set of locations L is called an *anchor* on LOC relative to an utterance d, c if

$$g(IND.0) = l_d$$
$$<, g(IND.1), l_d; 1$$

where l_d is the discourse location determined by d and $<$ is the relation of temporal precedence on L.

$SIT.\Phi$ will be used to give the *meaning* of the sentence Φ as a relation between the utterance situation d, c and the described situation s. We write this relation as

$$d, c[\![SIT.\Phi]\!]s.$$

In the example this relation holds if and only if there exists an anchor g on $SIT.\Phi.LOC$ relative to d, c such that

$$in\ s : at\ g(IND.1) : c(marry), c(John), c(Jane); 1$$

Observe that the speaker's connection is a map defined on the (morphological) parts of the expression Φ and with values in the appropriate domains of the semantical model, i. e. $c(marry) \in R$, $c(John)$ and $c(Jane)$ are elements of D. This means that we in this example have adopted a rather simple-minded treatment of names: The map c picks out a unique referent in the described situation.

There is still one missing part of the story: How to compute in a principled way the situation schema $SIT.\Phi$ from the given sentence Φ?

The computation will be arranged in two steps. First we assign a simple context-free phrase structure to the given sentence, next we introduce a set of constraint equations which partly are associated with the nodes of the tree and partly with the lexical items.

To treat our sample sentence we consider the following context-free grammar:

$$
\begin{array}{cccc}
S & \longrightarrow & NP & VP \\
 & & (\uparrow ARG.1) =\downarrow & \uparrow =\downarrow \\
\\
VP & \longrightarrow & V & NP \\
 & & & (\uparrow ARG.2) =\downarrow
\end{array}
$$

Using this grammar our sentence can be assigned the following structure:

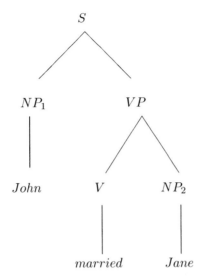

In addition to the "constraint equations" introduced by the syntactic rules, the lexicon adds a further set. In our case we have the following lexicon:

John	$NPROP$	$(\downarrow IND) = John$
Jane	$NPROP$	$(\downarrow IND) = Jane$
married	V	$(\uparrow REL) = marry$
		$(\uparrow LOC) = \downarrow$
		$(\uparrow POL) = 1$
		$(\downarrow IND) = IND.1$
		$(\downarrow COND_{loc}REL_{loc}) = precede$
		$(\downarrow COND_{loc}ARG'.1) = (\downarrow IND)$
		$(\downarrow COND_{loc}ARG'.2) = IND.0$

The equations from the syntactic rules and the lexicon must be combined to produce the associated situation schema. The equations are general and contain a set of "meta-variables", the arrows $\uparrow$ and $\downarrow$, which must be instantiated in each particular application. The procedure is as follows: Let f_1 be a functional variable assigned to the top node S of the tree. With some of the other nodes of the tree there will be an associated equation derived from one of the syntactic rules, e. g. with the node NP_1 we have the equation $(\uparrow ARG.1) = \downarrow$. The $\uparrow$-arrow of this equation points to the mother of this node, i. e. to the node S, and the

↑-arrow shall be replaced by the variable f_1. The ↓-arrow points to the node under consideration, i. e. the NP_1 node. We must then introduce a new functional variable f_2 to replace ↓, and we get the equation

$$NP_1: \qquad (f_1 ARG.1) = f_2$$

In the same way we get the equations

$$
\begin{aligned}
VP: & \qquad f_1 = f_3 \\
NP_2: & \qquad (f_3 ARG.2) = f_4
\end{aligned}
$$

The lexicon contributes a new set of equations:

$John:$ $\quad (f_2 IND) = John$

$Jane:$ $\quad (f_4 IND) = Jane$

$married:$ $\quad (f_3 REL) = marry \qquad (f_5 COND_{loc} REL_{loc}) = precede$
$\qquad\qquad (f_3 LOC) = f_5 \qquad\quad (f_5 COND_{loc} ARG'.1) = (f_5 IND)$
$\qquad\qquad (f_3 POL) = 1 \qquad\quad (f_5 COND_{loc} ARG'.2) = IND.0$
$\qquad\qquad (f_5 IND) = IND.1$

The situation schema is an attribute-value system. This means that our equations are functional equations. Thus an equation $(f_1 ARG.1) = f_2$ means that f_1 is a function which is to be defined on a domain including the attribute $ARG.1$, and the equation tells us that the value of f_1 applied to the argument $ARG.1$ is another function or attribute-value system. The correct interpretations of the last three equations under the entry $married$, is that f_5 applied to the argument $COND_{loc}$ gives a new function as value and that this function is defined on a domain including the attributes REL_{loc}, $ARG'.1$ and $ARG'.2$. Our equations impose certain constraints on the "unknown" functions $f_1, \ldots, f_5$. When we seek a solution to this set of equations we shall always look for a minimal, consistent solution. *And this minimal, consistent solution will be the situation schema associated with the given sentence.* And the style of writing the schema is but one way of exhibiting finite functions in tabular form. For a further discussion of the relationship between attribute-value systems and logic see Fenstad et al. (1987), pp 127–130.

Remark. In a certain sense unification and type theory represent two different strategies on how to "glue" the basic components of the grammatical analysis together. They look different, but are in substance quite similar (not identical), representing different ways of distributing the "theoretical mass" over the construction. In type theory you tend to "freeze" such constructions in a very rigid way as objects inside the theory, whereas in the unification approach they are kept external as part of the general mathematical tool-box. A good example is the interpretation of situation schemata in systems of situation semantics, Fenstad et al. (1987), pp 68–76. Here we in an informal, but standard mathematical way make use of a number of type-theoretical constructions, which in a pure type-theoretic approach reappear as objects inside the system.

Linguists generally recognize three main representational modules in the description of language structure, the phonological structure, the syntactic structure, and the semantic/conceptual structure. The disagreement comes when we ask about the inner structure and relative autonomy of the different parts and how the parts are bound together. I have in this section presented one point of view. But there are, indeed, many different stories being told. One version, which I in many ways find attractive, can be found in two recent books *Foundation of Language* by R. Jackendoff (2002) and *Simpler Syntax* by P. W. Culicover and R. Jackendoff (2005); see also the paper by Jackendoff in Brown and Hagoort (1999). Starting with Chomsky's *Aspects of the Theory of Syntax* Culicover and Jackendoff tell a story of how "mainstream" syntax has developed through various stages of the chomskian enterprise, including the current version of minimalism, ending in a final "flat" structure inspired by the attribute-value formalisms of Lexical-Functional Grammar (LFG), see Bresnan (1982, 2001), and Head- Driven Phrase Structure Grammar (HPSG), see Pollard and Sag (1987, 1994). This is a piece of "deconstruction" where the elaborate tree structures and transformations of chomskian syntax are transformed into a relational form, which – and this is a central claim of the authors – are more amenable to interacting with a semantic/conceptual structure.

It is instructive to compare the two stories. In both cases a somewhat rigid syntactic tree-structure (in one case minimalism, in the second case categorial grammar) is replaced by an LFG-like simpler structure. In both cases the simpler syntax is linked to a conceptual/semantic structure. And here is the point where the two accounts come together. The similarity between the CS (Conceptual Structures) of Culicover and Jackendoff (2005) and the situation schemata of Fenstad et al. (1985) is deep and immediate. Both are constraint-based formalisms and both allows for partiality, and their basic formats are quite similar.

A further technical remark. For the reader who is familiar with the notions of conceptual structures (CS) and situation schemata I add the following technical remark: The basic format of a CS, as introduced in Culicover and Jackendoff (2005), is:

$$\text{FUNCTION}(\text{ARG}_1, \ldots, \text{ARG}_i); \text{MOD}_1, \ldots, \text{MOD}_m;$$
$$\text{FEATURE}_1, \ldots, \text{FEATURE}_n.$$

The basic format of a situation schema, as introduced in Fenstad et al. (1985,1987), is:

$$\text{REL}, \text{ARG}_1, \ldots, \text{ARG}_n, \text{LOC}, \text{POL}..$$

We see the similarities, FUNCTION corresponds to REL; both formats have an ARG-list; FEATURES correspond to LOC. The MOD-list is missing from the situation schema format. This is because we at that time had the task to create a tailormade interface between LFG and situation semantics. In LFG modifiers are basically attached to either the REL, an ARG, or possibly the LOC attribute. Culicover and Jackendoff (2005) argue for a flatter syntax; see as an example the different treatment of NPs in LFG and in *Simpler Syntax*. The need for a semantic/conceptual representational interface is clear; its particular format will have to depend on your choice of syntax and semantic structure. We shall elaborate this issue further below, in particular, in connection with the theory of *Conceptual Spaces*, see Gärdenfors (2000).

Returning from technicalities to the main story let me conclude these brief remarks on grammatical structure by one further comment. In Jackendoff (2002) there is a fourth component, the spatial structure, see fig. 1.1 on page 6 of Jackendoff (2002). Precisely how this part is linked to the other components is not explained in any detail. In Fenstad et al. (1985, 1987) there is a well-defined fourth component, viz. the model structure or, in other words, a semantic/conceptual space. And there is a well-developed theory of interpreting situation schemata in the model structures; see Fenstad et al. (1987, pp 52–76).

1.4 Language technology

We have so far concentrated on the relationship between grammar and logic, i. e. on the theoretical understanding of how structure and meaning are connected. The new computer science has added new possibilities; suddenly there was a promise of 'true' engineering applications. 'Applied linguistics' existed before — there were some applications e.g. to speech disorder, to language teaching etc. — but applications were never a major concern of the community. Computers changed that.

The impetus came from an outsider. In 1949 Warren Weaver of the Rockefeller Foundation wrote a famous memorandum on the possibility of automatic translation. His first approach was to look at translation as a decoding task, taking his cue from the successful work on cryptography during the war. Add to this the possibilities of efficient large scale symbol manipulation provided by the emerging computer technology and you have a promising research agenda. It was soon realized that machine translation could be a useful technology in the cold war efforts, and suddenly money started to pour into an activity, which — to be honest — was never more than a promise.

One of the early actors in the translation enterprise was Y. Bar-Hillel. We met him above in our discussion of the categorial grammar of Ajdukiewicz. It was Bar-Hillel's idea that categorial grammar was a good instrument for the decoding task, but the early hopes came soon to an end — the complexity of language structure goes beyond what can be captured by categorial rules. It is fascinating to follow the early history of machine translation through a series of reports written by Bar-Hillel, tracing the development from the optimistic visions of his first report on *The State of Machine Translation* from 1951 to the pessimistic appendix *A Demonstration of the Nonfeasibility of Fully Automatic High Quality Translation* from 1960; see the reprint collection Bar-Hillel (1964). The pessimism of the 1960s almost killed machine translation, although some smaller communities survived. Bar-Hillel himself left the field and became a strong supporter of the Chomskian revolution of the late 1950s. One key innovation of *Syntactic Structures* was the concept of transformation. Adding transformations, such as the active-passive transformation, added power to Chomsky's analysis, but one should note that transformations are computationally rather intractable; indeed, there were for a long time no computational praxis as part of the Chomskian enterprise.

A new lease on life for machine translation came with the European Commission's large scale attack on the problem through the EUROTRA project of the 1980s. There were also some smaller projects in the US and in Japan, but this is not the place for a comprehensive history of machine translation. It is easy to understand why the Commission should be interested in the development of systems for automated translation, and it is equally easy to understand the willingness of linguists, both theoretical and computational, to participate. But once more one was chasing a vision without the proper foundation in a technology — EUROTRA was never a success.

However, machine translation with more limited aims has survived as a useful technology. Texts and the manipulation of texts are an integral part of computer technology, and we have witnessed, sometimes in an *ad hoc* fashion, the emergence of a new speciality, *computational linguistics*. Applications of this speciality did not always proceed according to the book of linguistic theory; for limited tasks a mastery of the basic technology and skill in programming were more important than adherence to the fine points of theory. As earlier remarked current systems for speech recognition make almost no use of linguistic theory. This should not come as a surprise, there is no difference in the relationship between linguistic theory and related engineering practices than we see otherwise in the relationship between a basic science and engineering applications. The interplay between science and technology is complex and not always well understood; there is seldom a royal and straight road from basic research to successful technologies in the marked place. We have today seen many useful applications of linguistic technology in the information sciences; but it is important to keep in mind that language technology, as any other technology, must live by its successes in the market place.

Computational linguistics, as we have noted, never was and is not today an applied branch of theoretical linguistics. There are many useful links and the equation "model = data base" is a useful starting point, but computational linguistics need to draw upon many sources; see e. g. the well-regarded textbook *Speech and Language Processing* (D. Jurafsky and J. H. Martin, 2000). As a current example let me mention the KUNSTI project which is a large scale research program in computational linguistics supported by the Norwegian Research Council in the years 2002-2006, where KUNSTI is a Norwegian acronym for "Knowledge Generation for Norwegian Language Technology"; see B. Maegaard et al. (2006). Of particular interest from our point of view is the machine translation project LOGON, see Oepen et al. (2004, 2007). The aim is to construct a Norwegian-English system of text translation. In its initial stages the project was based on theories of semantic transfer of the types described above, using LFG as the input grammar and HPSG for the output generation. In the semantic transfer part the system was based on the Minimal Recursion Semantics (MRS) system of Copestake et al. (1995), which can be seen as an alternative to the use of situation schemata in Dyvik (1993). But problems with lexicography led in later stages to a hybrid system, combining linguistic theory with several stochastic techniques, see Oepen et al. (2007).

However, problems in lexicography go beyond purely statistical and computational techniques. A useful review of current work is given in Blutner (2002), *Lexical semantics and pragmatics*. After discussing the defects of current theories Blutner points to two ways out. One way is represented by a theory of the *generative lexicon* as developed by Pustejovsky (1995). This approach aims to enrich the lexicon with new generative mechanisms while staying as close as possible to current linguistic technology. The advantage of this is that it keeps the changes close to the existing computable structures and is therefore computationally tractable. Pustejovski's presentation of his theory is not always easy to understand and leaves some doubts to the technical coherence of parts of his work. In a recent Oslo thesis, *Leksikalsk semantikk*, K. Skrindo (2001) has presented part of the theory in a consistent way within the framework of *Situations, Language and Logic* (1987). Skrindo's thesis, written in Norwegian, also gives a useful review of much current work.

Blutner argues against the generative position adopted by Pustejovski and presents, as his second way, a pragmatic alternative, pointing to the need for mechanisms of contextual enrichment (i. e. a pragmatic strengthening based on contextual and encyclopedic knowledge). But much remains to be done to translate this into a workable theory and efficient computational praxis. In an early attempt Blutner made use of the semantic information theory proposed by R. Carnap and Y. Bar-Hillel (1953). This is noteworthy, but the model theory underlying this approach is rather poor in content and structure. Blutner has since used optimality theory in his lexicographical studies, see Blutner (2004a). But there is another way to enrich the standard "list structure" of logical model theory, by adding geometry to the predominantly algebraic and combinatorial structure of lists.

1.5 Symbols carry meaning

We have noted some limits in the relationship between the science of language and language technology. There are also major challenges in the relationship between the science of language and the broader cognitive sciences. Formal semantics must proceed beyond the equation "model = data base" in order to serve as a link between language and mind. The challenge is to understand how language and cognition are rooted in the behaviour of large and complex assemblies of nerve cells in the brain. Much of current linguistic theory proceeds at the level of symbol manipulation, such as sorting, ordering and comparing symbols. But we need to proceed beyond this level. A first step is a geometerization of model theory as a basis for a phenomenological model of mind, this is the theme of chapter 2. The next is to understand how the dynamics of brain cell interaction generates this geometry, this is the theme of chapter 3. Much is already known, but much remains before we have a firm understanding of this double task.

2

Grammar and Geometry

Standard theory of grammar postulates the existence of two modules, one being a conceptual module which includes what is often referred to as knowledge of the world, one being a computational module which is concerned with the constraints on our actual organization of discrete units, such as morphemes and words, into phrases. Much of current theory is a theory of the syntax/semantics interface, i.e. a theory of how to connect grammatical space (the computational module) with semantical space (the conceptual module). In addition there has always been much work on the structure of grammatical space. However, remarkably little work has been devoted to the structure of semantical space. Even the Montague grammarians rarely make any use of the structure of their models; it is almost always possible to stay at the level of lambda-terms.

Let me proceed to fill in a few details in this account. A major part of current linguistic theory has been focused on the investigation of grammatical space; see chapter 1. As to the nature of the conceptual component I shall at this point only make one preliminary assumption, that it has — at some level — a standard model-theoretical core. This claim is not uncontroversial; it is one of the aims of this chapter to argue that model theory — correctly understood — is a necessary link between grammar and mind. We shall explain why.

2.1 Formal semantics and its ontology

First order logic — seen from a proof-theoretic perspective — is a system of remarkable strength. It has a complete proof procedure, i. e. every universally valid formula is provable. And all of mathematics, dressed in its set theoretic garb, is formalizable within the system. Yet this strength is illusory; we shall explain why.

Logic, as well as language in general, has two sides; one is syntax and proofs, the other is semantics and validity. In first order logic we seem to have a perfect balance between the two. On the syntactical side we have notions such as proof and theorem, on the semantical side we have the notions of model and validity. These notions are in the general theory of first order logic perfectly matched through the celebrated Gödel completeness theorem: a formula is provable, i. e. is a theorem of first order logic, if and only if it is universally valid, i. e. true in all models. Gödel's theorem is primarily a technical result, but it also is an insight with broader explanatory power.

In a sufficiently wide sense, logic is a natural science: We have a pre-formal idea of truth and, hence, of a correct argument. Logic converts this preformal idea into two distinct technical notions, the syntactical notion of proof and the semantical notion of validity in a model. Informally we may convince ourselves that what is provable is necessarily true in the preformal sense, and what is true in the preformal sense is, in particular, true in all models. Thus the preformal notion of truth is caught between the two technical notions of proof and validity. Gödel's completeness theorem, which asserts that validity in all models implies provability, closes the circle and seems to show that the preformal notion of truth has a correct and adequate analysis in terms of the technical notions of proof and validity in all models. This is an exemplary piece of applied science; a natural phenomenon has been given a sound and comprehensive theoretical analysis. We can even go one step further. The notion of proof, being a finite combinatorial structure, is algorithmic. The notion of being provable is not: a formula is provable if and only if it has a proof — but we cannot effectively decide in general if a proof exists or not. However, fragments of logic have effective proof procedures; it is a valid scientific strategy to explore the limits of effective computability and to combine the search for algorithms with various heuristic and probabilistic recipes. From this perspective reasoning and understanding seems to be reduced to the search for ever more sophisticated proof procedures. But, despite the successes, the strength is illusory.

The reason is ontological. The true ontology of first order logic is an ontology of lists; we explain why: A model for first order logic is a set-theoretic structure consisting of a non-empty set, the domain A, and a collection of relations, $R_1, R_2, \ldots$, defined on the domain A. We use the word set-theoretic to emphasize that an n-ary relation at this level of explanation is nothing more than a set of n-tuples over the domain. An n-ary relation R on A can be represented in the form of two lists.

One a list of basic or atomic positive facts, i. e. a list of all

(2.1) $$Ra_1 \ldots a_n$$

such that $(a_1, \ldots, a_n)$ belongs to the relation R; and a supplementary list of basic or atomic negative facts,

(2.2) $$not\text{-}Rb_1 \ldots b_n$$

for $(b_1, \ldots, b_n)$ not belonging to the relation R. The lists which may be infinite if the domain A is infinite, are complete, i. e. any n-tuple over A defines an item in either list (1) or (2). The important thing to recognize is that the model structure is "flat". All objects of A has the same ontological status, there are no structurally defined hierarchies. Of course, a relation R may in a specific structure impose a hierarchy on its domain. But to prove deductive completeness theorems we must necessarily speak of all models. This means, e. g. in applications to set theory, that the semantical interpretation is a coding of a possibly rich structure into a flat domain. This may be adequate for book-keeping purposes, even for linguistic engineering applications, but not for understanding, in particular, for natural language understanding.

This limitation of first order logic has always been recognized, and several extensions have been studied. We mention briefly:

higher types and partiality,
possible worlds and situations,
order/time/events and episodic logic,
masses and plurals.

The literature on extensions of first order logic is vast; we can only mention a few references of special relevance for our current purposes. Partiality and higher types is reviewed in Fenstad (1996a). Situation theory, partiality and the relational theory of meaning is the topic of Barwise and Perry (1983). The theory of mass nouns and plurals have been studied by G. Link and J. T. Lønning; for general references see Lønning (1987 and 1989). The extensions are, however, "tame"; all of them rest on the ontology of lists. A partial model corresponds to partial lists, which means that if the relation R in a structure A is partial, then there are n-tuples over A which occur neither in (1) nor in (2) above. In a similar way structures of higher types correspond to lists of lists, and possible world structures correspond to collections of indexed lists. A small extension occurs with order and time. In these cases the domain A has a fixed order relation, but since theories of linear, partial, and other types of order are algebraic and easily axiomatized, we remain with the same ontology. We should be careful to note that this ontology is adequate and productive for many applications in language technology.

We have recently seen a fruitful merging of ideas from data base theory and logic. This is not the place to review current advances in relational and deductive data base theory and to see how data base theory is becoming more and more intertwined with finite model theory. We shall, however, indicate the connection by presenting a simple example of a Q-A (i. e. question-answer) system.

The system has the following architecture, Vestre (1987); see also the exposition in Fenstad, Langholm and Vestre (1992):

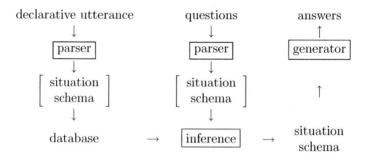

The system stores factual information in a data base. This data base can be updated by accepting input sentences within a natural language fragment. The parser rewrites such sentences in the format of a situation schemata, which is a particular kind of an attribute-value structures, Fenstad et al. (1985). The system automatically extracts from the atribute-value structures basic fact — positive or negative — and stores these facts in the data base. A question is rewritten as an incomplete situation schemata. The inference module accepts these incomplete structures and tries deductively to transform the incomplete sign into a complete attribute-value structure, using information from the data base as admissible hypotheses for the proof. Finally, the generator transforms the completed sign into a grammatically correct answer. The architecture of the system is open. The system can be extended to accept a larger class of sentences and questions by extending the syntactical component. The domain of application can be extended by reading into the system more information. The system can also be made more efficient by further development of the inference subpart. Seen in this perspective the system is successful as an example of language engineering. But there are limitations.

A sufficiently precise question will receive a correct and exhaustive answer. But if the question is more open and searching, it may signal that we are interested in answers which are "relevant" and "informative" rather than just "correct" in a strict logical sense. This may be the case when a question is part of a larger dialogue situation. An adequate analysis of dialogues is, however, not a matter of grammar alone. If this point is granted, we immediately see the limitations of the system, viz. the conceptual module, i. e. meaning and understanding, is reduced to a data base which is nothing more than a list of basic facts. The equation "model = data base" has useful technological applications, but it cannot serve as a basis for an analysis of the flow of meaning in a natural language dialogue.

To analyze the situation in some more details we need to review a few facts from how human language developed. This is a domain of many theories, often at odds with each other. What we need will only to a small degree be touched by these controversies. With some justification we shall therefore concentrate on only one account — the story as told in M. Donald, *Origin of the Modern Mind*, Donald (1990 and 1993). This is a story of cultural evolution. It needs to be supplemented by a study of the development of language in the child, a topic which we will return to in chapter 3.

According to Donald we can recognize four stages in the evolution of culture and cognition: *episodic culture, mimetic culture, mythic culture, and theoretical culture*. The episodic culture was characterized by the ability to react to a fixed situation in which the individual was placed. In a certain sense one "understood" the significance of this situation and could react accordingly. But this insight could not be used to understand when placed in other analogous situations. The transition to mimetic culture extended the limited mode of understanding found in episodic culture. The individual was now able to decompose the meaning of a specific situation into components and to recombine these in order to develop effective patterns of reaction in other, but similar situations. A rudimentary communication system between the members of a tribe also developed. According to Donald this was pre-linguistics and exclusively composed of gestures and other mimetic elements. The transition from mimetic to mythic culture was above all characterized by the acquisition of language. We now see an emerging communication system — a possibility for dialogue — which is a combination of gestures and speech. Donald is very careful to ground his theories in anthropological data on anatomical and cultural development. He stresses, in particular, how the development is cumulative, e.g. in a dialogue we have both mimesis and speech, the latter stage has not suppressed the former — a dialogue is not reducible to a text.

In the third transition, from mythic to theoretical culture, there are no anatomical changes. We see a transition to a stage where we have access to external symbolic storage. The development of written language was an early stage in this transition; present day information technology represents another. For the philosophically informed it may be instructive to compare Donald's analysis of the third transition to Popper's theory of the so-called World III, see Popper (1972); another and compatible story is the anthropological analysis of the concept of culture, see White (1947). There are many comments that one could want to make; I shall only emphasize one crucial new element occurring at stage four, viz. the possibility of an external memory without an anatomical foundation, or — expressed in a more colorful language — the possibility of a "collective mind" independent of the individual brains of the species. This explains why certain applications of linguistics to engineering are successful and why other fails.

A limited Q-A system can be completely located within the fourth cultural stage, i. e. theoretical culture with the possibility of external symbolic storage. None of the previous stages need to be involved; semantical space is therefore reducible to lists and data bases, as was the case in the system described above. But a real-life dialogue lives at the intersections of many cultures, in particular, the mimetic and the mythic. The dialogue is a combination of gestures and speech and is therefore not reducible to an external text. The dialogue is seen to pose other and more intractable challenges to the conceptual module; the successful analysis of dialogues, beyond a few stereotypes, is not a matter of incremental extensions of current linguistic technology. On the other hand, limited translation systems do belong to the fourth stage, which means that an ontology for translation is an ontology of external objects, and hence reducible to an ontology of lists and lexicons; for an interesting example of a limited and interactive translation system see the PONS system developed by Dyvik (1993).

The situation will be radically changed if we want to build a system for combined speech and vision. As a starting point we may be looking for something very simple — e. g. a system which shall recognize objects simply by name. To simplify even further the domain of application may be severely restricted, but we shall insist on a system which will model the task in the "correct" human way. This is a challenge almost totally within episodic culture, with a modest component of mythic culture, but with no component from mimetic culture. The task is, however, far from trivial — and not yet convincingly solved. The difficulty lies with the conceptual module. Meaning can no longer be reduced to a list or a data base; to succeed we need geometry. Current philosophy of language distinguishes three aspects of language, syntax, semantics and pragmatics which is also the current dogma of applied and computational linguistics (with the necessary addition of a phonological module to deal with speech). This is harmless for simple applications, such as ticket reservation systems, which operate entirely within the phonological-syntactic-semantic range. In more complicated applications, such as dialogue systems, current approaches foresee added pragmatic features — however, in an incremental way. We have suggested a different perspective based on a four-stage evolution of culture and cognition, where the stages are cumulative. Granted the correctness of this evolutionary history we would explain the success of limited systems, such as the PONS translation system, by the fact that such systems live almost totally within the fourth theoretical stage. In this perspective, "pragmatics" is not an addition to the syntax-semantics division in order to deal with a number of "rest factors", but a label which masks a number of radically different phenomena. There is a need for a new analysis of "pragmatics" within the broader context of a theory of cognitive development.

2.2 Model theory and geometry

In a historical perspective logic and geometry were partners from the beginning, the paradigm being the Euclidean axiomatization of geometry, a high point being the work of Pasch and Hilbert in the late part of the 19th century. But today logic and model theory seem to be in a much closer partnership to algebra and arithmetic. The reason for this has both historical and systematical explanations. From Descartes we saw a coordinatization of geometry; in the late nineteenth century we saw an arithmetization of analysis; Hilbert wanted to prove the consistency of geometry through a reduction to number theory; and Dedekind and Peano gave penetrating logical analyses of the systems of natural and real numbers. At the same time we saw the formalization of logic through the work of Frege, Russell and Whitehead. This line of development culminated in the first order formalization of set theory by Skolem around 1920. Thus everything conspired to give first order logic its prominence.

A model for first order logic is in essence an algebraic structure; and a fruitful partnership between algebra, arithmetic and logic has been established, starting from the Gödel completeness theorem. Natural axiomatizations of geometry and topology would seem to use higher order logic. Since such systems lack the compactness property with respect to the class of intended models, time was not ripe for further collaboration between logic and geometry. There was, however, some activity in the first order model theory of elementary geometry, see Tarski (1959). And a first start on a topological model theory was made by Flum and Ziegler (1980). But geometry was basically missing from model theory. Interestingly, the link between logic and geometry survived in other fields of study. One example is measurement theory. From the 1950s on there developed links between logic and foundational studies in measurement theory. The starting point was quite algebraic in spirit, scales being seen as systems with an ordering relation. But with multi-dimensional scales the geometric content necessarily becomes more prominent. Studies of perception and measurement led to the notion of a perceptual space, where the prime example is the theory of color space; for an account of these developments, see Suppes et al. (1989). Measurement theory is only one example of the "model-theoretic" approach to scientific structures developed by P. Suppes; for a comprehensive survey see Suppes (2002).

The need for a richer structure on the conceptual component was convincingly argued in the work on mental models by P. J. Johnson-Laird (1983). He was able to show how a simple geometrical representation of knowledge combined with the use of symmetry and invariance properties gave better (i. e. psychologically more plausible) models of reasoning than the standard deductions of formal logic. This insight was carried further by J. Barwise and J. Etchemendy in a paper from 1991, *Visual Information and Valid Reasoning.* In this paper we see a first presentation of the "hyperproof" program which is a system for combined visual and logical reasoning. This has sparked a rich development of systems for heterogeneous reasoning; for further references see Hammer (1995). We should also within the same circle of ideas mention the work of Habel on representation of spatial knowledge; see Habel (1990). He also argued for a dual coding for the processing of spatial expressions, with both a propositional and a depictorial representation. This is closely linked to the theory of mental models and ideas concerning heterogeneous reasoning. In all approaches the inspection of "mental models" combined with rule-based deductions are seen as essential ingredients in spatial reasoning.

We remarked above that even within Montague grammar, where the model theory seems to be given a prominent role, one almost never see in an actual analysis any geometrical structure. Most of the work can be carried out at the level of lambda-terms. But linguistic analysis sometimes forces one to pay attention to geometry; this was a point which I argued in an early paper on Montague grammar; see Fenstad (1978). We shall mention some examples. Our first example concerns the meaning of reciprocals; the geometry of "each other" is quite different in the two examples:

The men were hitting each other.
Five Boston pitchers sat alongside each other.

An interesting classification of the geometry of reciprocals, using the apparatus of generalized quantifiers, has been carried through by S. Peters and co-workers; see Dalrymple et al. (1994).

Locative prepositional phrases present another example where geometry is necessary for the semantical analysis; see E. Colban (1986). We cannot enter into the details of his theory but consider one simple example. In this case the prepositional phrase has an adjunct reading:

Peter ran to the school.

The semantical analysis proposed by Colban of this sentence refers a path or trajectory in space-time ending at the school, such that Peter is in the state of running on the curve. Colban's study presents a full fragment, connecting syntax and semantics through an implemented system.

Much of the work that we have reported on in this section can be conveniently brought under the notion of a *conceptual space* introduced by P. Gärdenfors in a number of recent papers; see Gärdenfors (1991, 1993 and 1994). A comprehensive treatment is found in Gärdenfors (2000). We conclude this section with a review of this work, showing in particular its connection to theories of natural kinds, to prototype theory, see Rosch (1978) and to cognitive grammar, see Langacker (1987 and 1991) and Lakoff (1987). The theory of *conceptual spaces* is an attempt to provide a theory of semantic structures suitable for linguistics and cognitive science. We noted above that standard model theory is basically a theory of lists. For technological applications, where the equation "model = data base" remains the operating modus, lists may well suffice – for cognitive science it does not. We have seen a refinement in situation semantics, where the location component, representing a connected region of space-time, plays an important role. But situation theory is very much a realistic theory. There is always a given *discourse situation* with a speaker, an addressee, an utterance and a location – and a *described situation* "out there", i.e. in a suitable sense "a situation in the world" (see chapter 5 of Gärdenfors (2000) for a more careful analysis). And the meaning relation in situation semantics is a complex relation between two situations and an utterance, the latter represented by a situation schema; see Barwise and Perry (1983) for an extended discussion. But even with this refinement of standard model theory, situation semantics is not exactly right for the analysis of concepts, mind and brain.

The starting point of Gärdenfors (2000) is the insight that *concepts* should be structured relative to several *domains*, which form clusters well separated from each other. *Color* and *shape* are typical examples of such domains. In the case of *color* we usually recognize three dimensions, hue, chromaticness and brightness – thus the *color* domain is a three dimensional space, where hue is represented by a circle, saturation and brightness are linear.

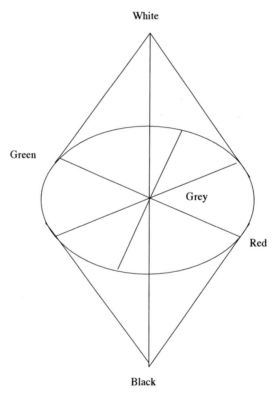

The dimensions determining a given conceptual space may either be inborn or culturally acquired. However this may be, the crucial fact is that each dimension comes equipped with a geometrical or, more generally, topological structure. In color space we have a metric structure defined by measuring length along the appropriate curve (a circle or a straight line). But conceptual spaces can be more abstract than perceptual spaces, one example is the two-dimensional conceptual space generated by the first two formants of vowels frequencies; see Gärdenfors (1991).

Concepts are usually related to several domains, "red cube" relates both to the *color* and the *shape* domains (and possibly to many others – what is the cube made of?). A *property*, following Gärdenfors (2000), is a concept related to one domain, e. g. "red" is related to the *color* domain only and can be identified with a subset of color-space.

In standard model theory properties are arbritary subsets – there is no further general analysis. This has caused philosophers endless difficulties in their attempts to give an analysis of notions such as "natural kinds" within the framework of standard logic. The added ingredient in the theory of conceptual spaces (taking a clue from the study of perception) is *geometry*.

A *natural property* is constructed as a convex subset of the model space. This is obviously the correct way in the case of colors; take any two points in the red sector of the color circle, then any point between is also red, i. e. red as a property is a convex subset of color space. More remarkably, the construction fits the facts in the conceptual space used for the phonetical identification of vowels. How far this assumption is valid, remains to be seen; we shall return to the point in our discussion of geometry and mind.

Here we point out one pleasing application, viz. a valid principle of induction for conceptual spaces: Let P be a natural property in the conceptual space S and let $a_1, \ldots, a_n$ be points of S belonging to P. Then the convex hull of $a_1, \ldots, a_n$, i. e. the least convex subset of S containing all points $a_1, \ldots, a_n$, is a subset of P; see the discussion in Gärdenfors (1994).

The proposal has also applications to theories of prototypes; see Rosch (1978). In traditional logic properties has been identified with subsets of the domain of interpretation. Granted no further structure there has been no end to the philosophical discussion of what are the "natural" properties and how do we determine when two objects share the same property — when is a tiger a tiger, what exactly are the necessary and sufficient conditions for a chair to be a chair? Prototype theory has developed as an alternative to the logical approach which was based on lists of necessary and sufficient conditions. Natural properties form convex sets in a suitable conceptual space, certain exemplars are more central or typical as examples of the property, i. e. they may serve as prototypes, and the extent of the concept is a convex neighborhood of the accepted prototypes; see Rosch (1978), Mervis and Rosch (1981), Lakoff (1987), Gärdenfors (1993), and Murphy (2002). Conversely, if we are given a set of categories and have decided on a set of proto-types $p_1, \ldots, p_n$ for them , then the prototypes determines the categories as a convex partition of the space:

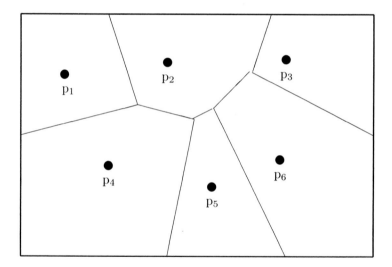

This partition procedure is an example of a so-called Voronoi tessellation; for a comprehensive review of this technique see Okabe et al. (1992). We shall continue this discussion in the next section, translating the discussion into the language of dynamic processes. Here it remains to add a few comments on cognitive grammar.

We expressed in the introduction to this paper some sympathy for the revolt of the cognitive grammarians. But we also noted with some surprise the rather primitive apparatus used by Langacker (1987) in his exposition of the basic ideas. P. Gärdenfors has in *Conceptual Spaces as a basis for Cognitive Semantics*, see Gärdenfors (1993) shown how model theory with a geometric extension may serve as a basis for cognitive grammar. He agrees that meaning is conceptualization in a cognitive model, not truth conditions in possible worlds. He notes that cognitive models are mainly perceptually determined and that semantical elements are based on spatial or topological objects. He argues in agreement with the cognitive grammarians that semantics is primary to syntax and that concepts show prototypical effects. All of this leads to the conclusion that conceptual spaces define a "correct" framework for cognitive grammar. We agree — but only up to a point.

The theory of conceptual spaces is not a "universal method" powerful enough to explain everything about meaning in human communication; see our discussion above of Donald's theory of the evolution of culture and cognition (Donald, 1993), in which he recognizes four stages: episodic culture, mimetic culture, mythic culture, and theoretical culture. It is in my view well established that some elementary parts of abstract computations are activities of the episodic/mimetic stages of development, thus activities of the individual mind and can as such be explained in terms of current brain models, e.g. neural networks, and current theories of mind, e.g. conceptual spaces. But "higher arithmetic" requires also the theoretical stage, which according to Donald is an external stage transcending the individual mind; in this connection see, in particular, the discussion in Donald (1993) of the changed role of biological memory in the transition from the earlier internal stages to the last and external theoretical stage.

The theory of conceptual spaces lies in the middle ground between language structure and brain dynamics. From the linguistic side we seem to have the technology available to connect a syntactic analysis in an LFG format to the semantics of conceptual spaces, using an attribute value formalism extending the approach used in connection with LFG and situation semantics – we note that refinements such as multi-dimensionality of domains and geometry are no serious technical obstruction. Explicit constructions, however, need to be supplied to turn this opinion into a solid fact. But far more challenging is the interface between conceptual level and actual brain. How is the geometry of conceptual spaces generated by an underlying brain dynamics?

2.3 Geometry and mind

We have so far made a journey from grammar via semantics to geometric structure. In this part we will start out with some remarks on brain dynamics and explain how the associated processes lead to certain geometric structure spaces. The main suggestion of this chapter is the proposal to identify the geometric structure space derived from brain dynamics with the geometric model theory discussed in the last section, the link being the identification of the notion of a "natural property" seen as a convex region of logical model space with the property of being a domain of attraction of an attractor of the brain dynamics; we shall briefly discuss both color space and prototype theory from this point of view to highlight the connection. A theory of mind would then be founded on the theory of this class of geometric structure spaces, and granted the identifications mentioned above, we would have a seamless connection from grammar via geometry to mind. Such are the ambitions; we shall see how far they are able to withstand the complexities of the real world.

The affirmative side of the story was presented in a review article, *Neuronal Models of Cognitive Functions*, by J. P. Changeux and S. Dehaene (1989). They give a survey of current work at the interface between cognitive science and neuroscience, hoping to build bridges rather than enlarge on differences. "The real issue becomes the specification of the relationship between a given cognitive function and a given physical organization of the human brain. From an experimental point of view, our working hypothesis (rather than philosophical commitment) is that levels of organization exists within the brain at which a type-to-type physical identity might be demonstrated with cognitive processes (op. cit.)". In their article they review, in particular, work done on short-term and long-term memory and the recognition, production and storage of time sequences. They deliberately pass by a discussion of complex cognitive functions such as problem solving and language processing, since at this stage of development of the theory the neurobiological basis is too complex and the relevant data too few for fruitful mathematical modeling activity. But this is at most a temporary restriction; their final conclusion is that "it is timely to approach cognition in a synthetic manner with the aim to relate a given cognitive function to its corresponding neural organization and activity state (op. cit.)". We shall have more to say about the strong assumption of one-to-one correspondence between cognition and brain processes as claimed by Changeux and Dehaene. But first we shall briefly recall another, but related approach to cognitive modeling.

Our next example is the theory of *attractor neural networks* developed by D. J. Amit; see his book *Modeling Brain Function*, Amit (1989). An attractor neural network consists of a finite set N of nodes or neurons, $a_1, \ldots, a_N$ each of which can in the simplest case be in one of two possible states, $s_i = 0$ or $s_i = 1$. Each pair (a_i, a_j) of neurons is connected through a function J_{ij} which measures the influence or the synaptic efficacy which the node a_i may have on the node a_j. We assume that the process is a discrete time process; its dynamics will be given by the two equations:

(2.3)
$$s_i(t+1) = ch(h_i(t+1) - T_i)$$

(2.4)
$$h_i(t+1) = J_{i1} s_1(t) + \cdots + J_{iN} s_N(t)$$

The system operates as follows: At time t the system is in stage $s = (s_1, \ldots, s_N)$. At each node there is defined a local field h_i. This local field is at stage $t+1$ determined by the current stage s and the channel weight functions J_{ij} according to equation (4). The function T_i represents a threshold level at node a_i. If the field h_i at a_i at time $t+1$ is greater than T_i, then neuron a_i fires, i. e. the value of $s_i(t+1)$ is set to 1; if h_i is less than T_i, then neuron a_i is inactive, i. e. the value of $s_i(t+1)$ is set to 0. The function ch formalizes this description, i. e. $ch(n)$ is 1 or 0, depending upon whether n is positive or negative; this is the content of equation (3).

The dynamic behavior of such systems can be quite complicated; we recommend Amit's book as an excellent guide to the field. The analogy to spin-glass theory in non-linear statistical mechanics is striking. Assuming full connectivity, i. e. J_{ij} is defined for each pair i, j, and symmetry, i. e. the validity of the set of equations $J_{ij} = J_{ji}$, the full force of mean field theory comes into play and it is possible to analyze the dynamics of an attractor neural network in great details, in particular, to give a rather full description of the set of attractors of the system. From a neurobiological point of view, the assumption of full connectivity and symmetry is suspicious, if not outright false. But as Amit argues, they form a convenient starting point; through some further "robustness studies" it is also possible to see how the assumptions may be relaxed in order to obtain models more faithful to neurobiological fact. And, indeed, current research seems to suggest this view. The brain can to some degree be viewed as a family of networks, each being to some extent fully connected, whereas the family is connected in a much "looser" sense. The reader should be aware of the fact that there is much mathematics behind this remark, see the section *Complex network of networks for understanding brain dynamics* in J. Kurths et al. (2009).

For the moment let us play with the simplified version of an attractor neural network. It is easy to see how a system defined by the equations in (3) and (4) can be made memorize certain patterns: In fact, let $p^1, \ldots, p^K$ be K prescribed patterns, i.e. each p^k is an N-tuple of 0's and 1's, $p^k = (p^k_1, \ldots, p^k_N)$. Define the synaptic connectivity by the equations:

$$(2.5) \qquad J_{ij} = (p^1_i p^1_j + \cdots + p^K_i p^K_j)/N$$

Then a simple argument shows that the patterns $p^1, \ldots, p^K$ are the fixed point attractors of the dynamics defined by the set of equations (3), (4) and (5). With all our simplifying assumptions the system also has an energy function defining a suitable geometric phase space of the system. The configurations p^k, being the attractors of the system, determines both the local minimum locations on the energy surface, as well as their domain of attraction It is tempting to see such geometric space as the proper starting point for a phenomenological theory of mind.

In the previous section we saw how the notion of a color space could be reconstructed as a conceptual space. Colors correspond to convex regions in the space; let us for the moment think of hues only, i. e. focus on the color circle. According to E. Rosch colors have prototypical properties, thus the geometry of color space is determined by a Voronoi tessellation based on a finite set of prototypical exemplars. In a similar way color prototypes can be interpreted as a fixed set of patterns to be stored by a suitable attractor neural network; for a related discussion of prototypes and neural networks see chapter 11 of J. A. Anderson (1995). In this case the prototypes are the attractors of the system and the concept of color corresponds to a domain of attraction in the energy surface of the system. This would give a reasonable dynamics for color perception. The correspondence between convex geometry and the dynamics of attractors is quite close; granted sufficient regularity assumptions the claim is that the two accounts tell basically the same story. In this way we see a connection between grammar and mind — the link being geometry.

The storage and retrieval of temporal sequences is a first important step beyond the "passive" networks described by equations (3) and (4) above. D. J. Amit has constructed an attractor neural network that is able (in a precise technical sense) to count chimes; see Amit (1989). Amit sees this as a tentative step into abstract computation. His suggestion should be reviewed in the context of a theory of cognitive development. Recognizing and reacting to a fixed number of chimes may be an activity of episodic mind only. But theoretical arithmetic, e. g. the recent proof of Fermat's theorem, is beyond any doubts an activity of the fourth stage of cognitive development. It is therefore a questionable exercise to attempt an account of mathematics in terms of neural networks, i. e. as an activity of an individual mind. The act of doing mathematics is rather a subtly shared activity involving several cognitive stages; see the discussion on the changed role of biological memory in Donald (1990).

Returning to our main theme we would like the picture drawn by Amit to be true, and it may, in principle, be so as an account of individual mind. But what we today know about actual cognitive processes in the brain, tells us that the simple and sometimes one-to-one connection between cognition and neuronal activities postulated by current work in neural network theory is currently far off the mark. Let me briefly point to some of the facts. On the positive side we have the wealth of results obtained through such techniques as PET scanning; see e. g. Kosslyn and Koenig (1992) and Posner and Raichle (1994). This has made possible quite detailed models of cognitive functions, models which are well grounded in anatomical facts. But they also points to the complexity of the link between cognition and anatomy. One striking example is a study of lexical access which is reported in Posner and Raichle (1994); see chapter five of their book. In this chapter they study a hierarchy which starts with the act of passively viewing displayed words, through the gradually more complex tasks of listening, speaking and generating words. Each task in this hierarchical experiment is seen to activate distinct set of brain areas. The information is precise but true understanding is complex and still far away. I shall only make two comments: PET studies do identify brain areas active in a given cognitive task, but they tell us little about the mechanism involved. Typically, Kosslyn and Koenig has in their book a chapter on computations in the brain, but little use is made of this chapter in later parts of the book — simply because the "true" computational structures are not known; for a reasonably current review see Churchland and Sejnowski (1992) and Churchland (2002). My second comment is that while there certainly is a correspondence between cognitive function and neural structure and activity states, the correspondence is not necessarily one-to-one; different activity areas with different architectures may generate similar geometries.

Both reasons argue for an independent phenomenological theory of mind. This means a study of the geometry without presupposing a detailed knowledge of the underlying dynamic behavior. This is not an unusual situation in science; to mention one example we know that we in equilibrium thermodynamics are — at least in principle — able to reduce the phenomenological theory of heat to molecular motion, but that we in non-equilibrium theory are still largely ignorant of the dynamics and therefore must introduce separate equations for the phenomenological level. A beautiful example of this strategy is found in A. Turing's study of the chemical basis of morphogenesis, see Turing (1952). He assumed the existence of two active chemical substances — so-called concentrations of morphogenes in his language — which generate a non-linear dynamics governed by a pair of coupled diffusion-reaction equations. Within this model he was able to show how the geometry generated by the process could explain certain morphological phenomena, one example being the process of gastrulation. One reason for mentioning this work is that Turing's model is not only a beautiful example in itself, but that it may also teach a lesson how to model the connection between geometry and mind; for a recent exposition on pattern formation and diffuson-reaction equations see Meinhardt (1995).

This modeling task was attempted by R. Thom around 1970; see his papers *Topologie et Linguistique*, Thom (1970), and *Langage et Catastrophes: Eléments pour une Sémantique Topologique*, Thom (1973). The geometric locus for Thom is also an "energy surface" which is supposed to be derived from an underlying brain dynamics. Thom, however, does not explicate the dynamics. His discussion proceeds at a purely geometrical or phenomenological level. In the 1970 paper he classifies spatiotemporal verb phrases in terms of singularities in the energy surface, and derives a "natural classification" of such verb phrases in terms of his classification of singularities into seven classes. This should be seen in connection with the classification of "natural kinds" in terms of convex regions and domains of attraction which was discussed above. In the second paper he develops a more systematic discussion. It would be beyond the scope of this paper to survey this work here; see the exposition in Petitot (1995). Suffice it to say that he claims, in complete agreement with our previous discussion, that a noun phrase is described by a potential well in the dynamics of mental activities and a verb phrase by an oscillator in the unfolding space of a spatial catastrophe. However, Thom's work has not had the influence on theoretical linguistics which it merits. There seems to have been an incompatibility of minds. Thom severely criticized Chomsky for his combinatorial approach as being totally inadequate as a theory of linguistic meaning. Linguists, unfamiliar with the mathematics, saw little relationship between Thom's "speculations" and their science. One early attempt to bridge the gap can be found in my paper on Montague grammar *Models for Natural Languages*, Fenstad (1978). I argued with reference to Thom's work for a link between him and Chomsky based on a geometrization of the model theory of Montague. But I did not pursue the topic further at that time. Today we see that there is a close connection between the theory of conceptual spaces, cognitive grammar, and the geometric approach of Thom, see the discussion in Gärdenfors (2000), Howard (2004), and Petitot (1995). It would also be useful to connect the study of conceptual spaces with the broader area of *pattern theory*, see the comprehensive review in U. Grenander and M. Miller (2007). I also strongly recommend D. Mumford's lecture *Pattern theory = a unifying perspective*, Mumford (1992); see also Hallinan et al. (1999).

In conclusion I would like to discuss some current work on sentence processing which fits into the picture drawn here. This is work done by G. Kempen and T. Vosse, see Kempen and Vosse (1989). They have developed an attribute-value approach to grammar called segment grammar:

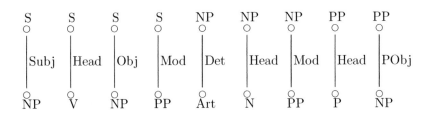

The grammar has two kinds of signs. One type - the so-called syntactic segments exhibited in the figure above - has a purely structural function. The other type - the lexical signs - are signs which carry linguistic meaning; see the following example taken from the same paper:

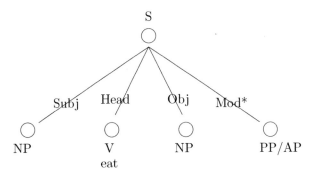

The authors introduce the notion of unification space, which to use their words is a kind of "test tube" filled with a mixture of syntactical and lexical signs. The structural signs are always assumed to be present in the tube; from time to time lexical signs are added and the tube well shaken. A reaction occurs and complete sentential signs are crystallized.

The pictorial language has a complete algorithmic interpretation. Signs combine through a unification procedure; see Kempen and Vosse (1990):

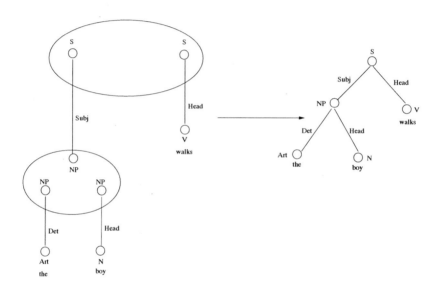

The dynamical behavior of the unification process is defined by a simulated annealing algorithm. Such algorithms, which have their root in statistical mechanics and material science, have become very popular in the field of combinatorial studies. In a sense they are very universal in nature, but not always efficient in execution. Since such algorithms are well known, we shall not give any details but rather suggest another possibility for the dynamics of unification spaces.

Let me recall an early attempt. In a number of works W. J. Freeman and co-workers at Berkeley suggested an alternative approach to neural modeling; for a convenient survey see Skarda and Freemann (1987 and 1990). Their particular concern is the olfactory system, but the aim is a general model of brain dynamics. They — of course — recognize a neuronal level of description but claim that it is not the right level for a general theory of brain dynamics. They find a philosophical alibi in the work of the French phenomenologist Merleau-Ponty and describes the dynamics as examples of non-linear and, hence, self-organizing systems; they could as well have made a reference to the work of Turing (1952); see also the current article, *The Brain, a Complex Self-organizing System*, Singer (2009). They argue for a chaotic dynamics, in particular, in order to account for the speed in the recognition of smells. In the Behavioral and Brain Sciences paper, Skarda and Freeman (1987), there is the usual critical review section. There may be critical comments, but the validity of a description at a phenomenological level cannot be denied. This was a point we also argued above. The complexity of brain mechanisms in the execution of cognitive tasks is so intricate that we are in any case forced back to a phenomenological level of description.

We have in this chapter traced a line from grammar to mind, the link being geometric model theory. Current computational linguistics concentrate to a large extent on the syntax/semantics interface. In most of this work the meaning component — the conceptual module — is reduced to a data base. This is adequate as long as we restrict attention to the final and theoretical stage of cognitive development, which is characterized by external symbolic storage, and which forms the base for current developments in language engineering. But it is not adequate as a basis for a theory of linguistic behavior in human beings. This is why we have insisted on including all stages of cognitive development in our account in order to appreciate the complexities involved. This perspective led to a geometrization of model theory as a basis for a phenomenological model of mind.

But mind could also be the starting point in our quest for understanding the connection between language and meaning. Our working hypothesis would then be that a phenomenological theory of "mind" is nothing but a branch of geometric model theory. The geometry is the basic object which introduces the lexical items, the rest — grammar, logic and mathematics — are formal tools used in the study of its structure. In particular, it seems possible to take a modified notion of unification space as a starting point for an account of language and speech, but in order to do so we need to explore a wider range of possible dynamics going beyond the simulated annealing algorithm used by Kempen and Vosse (1990). But this necessarily leads us into the "uncertain middle ground" between grammar and brain, which is the topic of the next chapter.

3

Grammar, Geometry, and Brain

Our approach has so far been top-down. We started in chapter 1 with a review of the formal theory of grammar, including some current theories of formal semantics. We then argued in chapter 2 for the necessity of enriching the formal semantics of logic with geometry in order to obtain a more satisfactory theory of meaning for natural languages. We concluded that chapter, perhaps a bit too boldly, with the claim that a phenomenological theory of meaning, and hence a theory of "mind", is nothing but a branch of geometric model theory. But there is no mind without a brain, and there can be little understanding of meaning and human communication without an insight into how the mind and brain are linked. We have so far told the story as seen from the top level of language and meaning. At the other end the neuroscientist has now command of a vast and extremely detailed knowledge of brain structure. But when the linguist try to reach deeper into the mind and the brain and the neuroscientist try to explain how higher cognitive functions emerge out of bare structure, we are in a somewhat uncertain middle ground. An exploration of this territory will be the topic of this chapter.

Language, meaning and brain is a rich and lively research area. There are many levels to explore and even more theories offered as explanations. No one can claim complete command of the field. But even the occasional traveller is struck by the lack of interaction between the participants in this research enterprise. Thus a secondary concern of this account is to point out possible connections between different approaches and how this may lead to theories of greater strength and applicability.

3.1 Brain and grammar

This section will be divided into two parts. If we want to understand how language emerged as a property of the brain, we are, on the one side, well advised to look at human evolution, how language developed in the species, as well as, on the other side, to focus on our current understanding of the structure of the brain and try to determine the neural correlates of higher cognitive functions.

Evolution. Few would doubt that an understanding of language development would facilitate our understanding of basic language mechanisms and how they are grounded in brain structure and functioning. Language today is the product of a long evolutionary development, and much may have become pre- or hard-wired in the brain during this development, obscuring the first stages in the development of speech, gestures and communication. And since it is an insight into the early stages of this development that most probably holds the clues to an understanding of how speech and meaning are connected, we seem to be in an impossible situation. Language does not leave any fossil records, thus the field is wide open to speculations and more or less "plausible" story telling. We could try to seek for clues in the early development of child language, but the study of how children acquire language is no sure guide to an understanding of the early relationship between sound and meaning, ontogenesis does not necessarily provide a blueprint for phylogenesis. But this does not mean that we in our quest for understanding the link between mind and brain should neglect the study of how children acquire language. The reader may wish to consult two recent surveys, *Infant pathways to language*, Colombo et al. (2009), and *How does early brain organization promote language acquisition in humans?*, Dehaene-Lambertz et al. (2008). The reader will find it instructive to compare the two accounts, the first being to a large extent based on statistical learning theory, the second founded on neural network models; for a broader review of learning theory see Meltzoff et al. (2009). We have chosen to focus on some evolutionary approaches.

Post Chomsky the study of language origin and development, guided by significant theoretical developments in both linguistics and cognitive science, have again become respectable; see e.g. the collections *Language Evolution* edited by Christiansen and Kirby (2003), *Language Origins* edited by Tallerman (2005), and the recent survey *The Origin of Meaning* by Hurford (2007). The reader may also wish to consult chapter 8, *An Evolutionary Perspective on the Architecture*, of Jackendoff (2002), which, in my view, presents a fair review of many of the issues involved.

One highly contentious issue in the study of language development concerns the status of language as a mental phenomenon: Is syntax a unique and autonomous feature of the human mind, or is syntax in some sense an emergent characteristics of the development of semantic complexity? This is usually referred as the problem of Universal Grammar (UG). One characterization of UG reads as follows; see Pinker (1994):

> **Universal Grammar.** The basic design underlying the grammar of all human languages; also refers to the circuitry in children's brain that allows them to learn the grammar of their parent's language.

In the Chomskian tradition, universal grammar is a specific and independent "language organ", genetically hardwired into the brain of children. But there are also many other plausible stories one could tell. In addition to the references above, see also P. T. Schoenemann (1999), *Syntax as an emergent characteristic of the evolution of semantic complexity* and chapter 13, *Language evolution*, in M. Nowak (2006).

Models of how language developed can be constructed on many levels. In addition to networks and statistical learning theory we have also some very abstract and formal versions. One example is J. Goldsmith (2007), *Towards a new empiricism*. He takes a step back from neurons and mental life, in fact, his starting point is family of grammars G_n based on a common alphabet, and according to him the task of Universal Grammar is, given a set D of data, to choose the "best" grammar accounting for the data. To do this he uses Bayes' theorem

$$\Pr(G|D) = c_D \Pr(D|G) \Pr(G),$$

where c_D is a constant depending only on D. To obtain an expression for $\Pr(D|G)$ he suggests the use of the theory of probabilistic grammars proposed by Solomonoff (1997). And he indicates how the complexity theory of Kolmogorov can be used to calculate the measure $\Pr(G)$. The aim of UG is then to find a grammar G* such that

$$G^* = \operatorname{argmax}_G \Pr(D|G) \Pr(G).$$

The setting is highly abstract, using e.g. the theory of Universal Turing machines to find an expression for the measure $\Pr(G)$. Neither man nor mind is involved, and it remains to be seen if this approach is a "new empiricism" which can be used to elucidate the notion of UG as a "human" phenomenon.

Let me now turn to another formal model proposed by Nowak and co-workers to explain language evolution; in addition to the book referred to above, see also the paper in *Science* by Nowak, Komarova and Niyogi (2001). According to their view UG consists of a mechanism to generate a search space of possible grammars and a learning procedure that specifies how to evaluate sample sentences. They claim, with reference to the so-called Gold's theorem on the non-existence of an algorithm for language learning, that both the mechanism and procedure must be partially innate. Gold's theorem is a result about formal language theory, and one may legitimately ask about its relevance for "real life", see both Jackendoff and Schoenemann for some divergent views. Jackendoff has a brief parenthetical reference in connection with learnability questions related to the *Aspect* model of Chomsky (Jackendoff (2002), p.77), but the theorem does not seem to play any important role in his subsequent discussion of UG. And Schoenemann dismisses in his paper the theorem outright as an improper application of a formal theorem to the "real" UG discussion. We shall let the reader judge the issue for himself and instead turn to an exposition of the formal model put forward by Nowak and co-workers. We want to answer the question, when does a universal grammar U induce coherent grammatical communication within a given population. The universal grammar U will generate a search space $G_1, G_2, \ldots, G_n$ of candidate grammars. Each grammar G_i is a rule system that defines a set of valid sentences. Let a_{ij} denote the probability that a speaker using G_i formulates a sentence also compatible with G_j (thus $a_{ii} = 1$). A basic assumption is that there is a reward in mutual understanding, and the appropriate payoff function is

$$F(G_i, G_j) = (1/2)(a_{ij} + a_{ji}).$$

Further let x_i be the frequency of individuals in the population who use G_i. The average payoff is given by $f_i = \sum_j x_j F(G_i, G_j)$. We also assume that payoff translates into reproductive success.

The second part of a universal grammar postulates a learning mechanism. Children learn from their parents, but learning can involve mistakes, let Q_{ij} denote the probability that a child learning from a parent using G_i ends with grammar G_j. With this notation in place the population dynamics is given by the equation

$$\dot{x}_i = \sum_{j=1,\ldots,n} x_j f_j Q_{ji} - m\, x_i,$$

where $i = 1, \ldots, n$ and $m = \sum_i x_i f_i$ is the average fitness or grammatical coherence of the population. We are interested in seeing how coherent communication evolves. We refer the reader to Nowak (2006) for the general theory and look here only at a simple example. First assume that each $a_{ij} = a$, for some constant a, when i differs from j. This will imply that each $Q_{ii} = q$, for some constant q, which is then the probability of learning the correct grammar. Next assume a simple model of learning, the "memoryless learner", in which the teacher (parent) uses a grammar G_k and the learner start with a randomly chosen grammar G_i. The teacher generates sentences consistent with G_k. As long as they also are consistent with G_i, the learner retains this grammar. If not, he chooses at random a new grammar G_j. At a certain stage the process will be stopped, let b be the number of sample sentences produces in the learning process. We shall omit the mathematics and just assert that it is possible to calculate two numbers q^* and C^*, such that if $q > q^*$ then the learning process will succeed as long as the sample size $b > C^* n$, note that the process is linear in the number of candidate grammars.

So much at this point for abstract mathematical models. In addition to the scientific issues it is also possible to look at the UG controversy as part of the history of how linguistics emerged as an independent academic profession. I have earlier in chapter 1 made some observations on Chomsky and mathematics. His early theories of syntax were influenced by the mathematical tools he had at his disposal, which above all was the theory of formal languages as part of general recursion theory. At that time model theory and formal semantics were not well established parts of mathematical logic. Great logicians such as Gödel and Skolem used in a deep and essential way a notion of structure in their work, but model theory as a separate discipline within logic emerged only in the mid 1950s as a byproduct of Tarski's great work on the notion of truth in formal languages. Thus formal semantics was not part of the tool-box available to Chomsky, hence syntax became the only possible solid foundation. And, as argued in chapter 1, the autonomy of syntax and the independence of linguistics as a pure science were two of the main characteristics of the Chomskian revolution in the study of language.

This brings us back to a model proposed by Nowak, Plotkin and Jansen (2000). In a paper in *Nature* some years ago they presented a model of how syntactic communication evolved. I quote from their abstract:

> Animal communication is typically non-syntactic, which means that signals refer to whole situations. Human language is syntactic, and signals consist of discrete components that have their own meaning. Syntax is a prerequisite for taking advantage of combinatorics, that is, "making infinite use of finite means". The vast expressive power of human language would be impossible without syntax, and the transition from non-syntactic to syntactic communication was an essential step in the evolution of human languages. We aim to understand the

evolutionary dynamics of this transition and to analyze how natural selection can guide it.

The setting of the model is necessarily simplified. Suppose a language of n words, $W_1, W_2, \ldots, W_n$, is given. An individual is born not knowing any of the words, but acquires words by learning from other individuals. With some simplifications the evolutionary dynamics is governed by the following equation, where x_i is the relative abundance of individuals in the population who know word W_i

$$\dot{x}_i = R_i x_i (1 - x_i) - x_i$$

where $i = 1, \ldots, n$. The product $x_i(1 - x_i)$ expresses the fact that the abundance of W_i spreads by the interaction of people who know the word with those who do not know it. The rate constant $R_i = bqf_i$ is the basic reproductive ratio of W_i, where b is the total number of word-learning events per individual per lifetime, q is the probability of memorizing a single word after one encounter, and f_i is the frequency of occurrence of W_i in the language. Finally, $-x_i$ expresses a certain constant death rate.

This framework is used to analyze how natural selection can guide the transition from non-syntactic to syntactic communication. Suppose we want to communicate about "events". Recalling from our discussion of situation semantics in chapter 1, we see that events in general may consist of situations or places, times, objects, and relations/actions. Nowak et al. simplify their discussion by assuming that events only consist of one object and one action, denoting by E_{ij} the event combining object i with action j. Non-syntactic communication uses word for events, syntactic communication uses word separately for objects and actions. It is further assumed that events occur at different rates, given by an event rate matrix, G. With this machinery in place the authors may define their main evolutionary concept, the fitness contribution, as the probability that two individuals know the correct word for a given event summed over all events and weighted with the rate of occurrence of these events. For non-syntactic communication the *fitness contribution*, F_n, is calculated to be

$$F_n = \sum_{ij} (x^*(W_{ij})^2) g_{ij},$$

here g_{ij} comes from the event rate matrix G, and $x^*(W_{ij}) = 1 - 1/R(W_{ij})$, where $R(W_{ij})$, the basic reproductive ratio of W_{ij}, is equal to bqf_{ij}, remembering that f_{ij} is the frequency of occurrence of event E_{ij}. A similar, but somewhat more complicated, expression is derived for F_s, the fitness contribution of syntactic communication; for details see the Methods section of Nowak et al. (2000). We only note that in decomposing the event E_{ij} into an object and an action, the single word W_{ij} is replaced by a combination of two words $N_i V_j$ (noun and verb), and similarly $R(W_{ij})$ is replaced by the pair $R(N_i) = (b/2)q_s f(N_i)$ and $R(V_j) = (b/2)q_s f(V_j)$, where q_s is the probability of memorizing a noun or a verb, and $f(N_i)$ and $f(V_j)$ denotes the frequency of noun N_i and verb V_j, respectively. The authors now run through a number of simulations to see when syntactic communication leads to a higher fitness than non-syntactic communication. In this discussion the following parameter p is of crucial importance. Suppose there are n objects and m actions; only a fraction of the mn possible events will occur in an actual world, let p denote this fraction. With $m = n$ (and granted in addition some further simplifying assumptions) one can show that syntactic communication has higher fitness than non-syntactic communication if

$$n > 3q/(pq_s).$$

Note that the parameter p expresses to some extent the compositional structure of the world. And we see that the smaller the value of p the larger n must be to satisfy the inequality, i.e. the more difficult it is for syntactic communication to develop.

To round off this discussion I include a few words on a recently proposed model, extending in some ways the approach of Nowak et al. discussed above. The model is presented in the paper *The consequences of Zipf's law for syntax and symbolic reference* by Ferrer i Cancho, Riordan and Bollobas (2005). They note, with reference to Nowak et al., that while the Nowak approach to the evolution of language formalizes why syntax is selectively advantageous compared with isolated signal communication systems, it does not explain how signals naturally combine. In their own study they start out with a simple model in which symbols are associated with objects. From this network they define another network, exhibiting a primitive form of word-word association, where two words are linked if they refer to at least one common object in the first network. Will this linking give any hint of how a "proto-language" could evolve? We quote from the abstract of the paper:

> ... recent work has shown that if a communication system maximizes communicative efficiency while minimizing the cost of communication, or if a communication system constrains ambiguity in a non-trivial way while a certain entropy is maximized, signal frequencies will be distributed according to Zipf's law. Here we show that such communication principles give rise not only to signals that have many traits in common with the linking words in real human languages, but also to a rudimentary sort of syntax and symbolic reference.

Returning to Nowak et al., we note that in their study a simple model of the world is the starting point, i.e. an understanding of the world is primary to speaking about the world. We should further note that world-object network in the Ferrer i Cancho et al. study also incorporates some world knowledge. Linking and syntactic decomposition are not arbitrary activities, but activities of the human mind in the world. This accords well with the assumption of a pre-existing primate conceptual structure, which is also the starting point in Jackendoff's discussion, see chapters 4 and 8 in his book referred to above, see also the discussion in Donald (1990). There is also a possible link to some recent studies of how children develops language, see *Language acquisition, domain specificity, and descent with modification*, Marcus and Rabaglati (2009). They have studied children's ability to extract rules from various linguistic and non-linguistic inputs and have found that "infants privilege speech in extracting rules from temporal streams ... We do not yet know why; infants may analyze speech more deeply than other signals because it is highly familiar or highly salient, because it is produced by humans, because it is inherently capable of bearing meaning, ... ". I would add to their search for an explanation: because of a possible pre-existing primate conceptual structure. The model proposed by Nowak et al., does not tell us anything about how humans evolved this elementary conceptual structure. It tells a story how some elementary form of "proto-syntax" can be the result of some evolutionary process governed by the equation exhibited above. This is a starting point, and from this starting point there are many plausible stories to tell. Gestures and meaning may have evolved before speech and syntax, but in the course of human evolution it may be hard, even impossible, to keep track of the detailed interaction between syntactic and semantic features and to account for what is learning and what has become pre- or hard-wired. And, given the close links between syntax and semantics, it is to be expected that some "linguistic fact" can be given either a syntactic or semantic explanation, as is well known from logic where one often can choose between a proof-theoretic or model-theoretic explanation. I shall not pursue these questions further, except to insist that conceptual structures, i.e. geometric model theory, must be part of the story, either as a driving force or, more possibly, as an equal partner in the complex co-evolution of meaning and speech.

The brain If we are to succeed in the task of explaining how meaning and mind are grounded in the physical brain, we first of all need detailed models of brain structure and functioning. This is a very active area of research. Much is now known about structures, less about functions.

Out of a vast literature let me mention a few general surveys which may be useful as a background to our reflections on grammar, geometry and brain: G. Marcus (2004), *The Birth of the Mind. How a Tiny Number of Genes Creates the Complexities of Human Thought*; P. Gärdenfors (2003), *How Homo became Sapiens. On the Evolution of Thinking*; and M. Donald (2001), *A Mind So Rare. The Evolution of Human Consciousness*. The books all report many facts and observations. They all try to weave this information into a coherent story connecting mind and brain. The stories may be plausible, but "hard" science it is not. Current research is regularly reviewed in journals such as *Nature* and *Science*. At regular intervals we have handbook-type comprehensive reviews, of special relevance is M. S. Gazzaniga (2004). The reader is also strongly advised to consult the critical review *23 Problems in Systems Neuroscience*, J. L. van Hemmen and T. Sejnowski (2006).

Closer to our immediate concern are the review articles in the book *The Neurocognition of Language*, C. M. Brown and P. Hagoort (1999). Of particular interest are the reviews in the section on the neurocognitive architecture of language, dealing with the basic brain architecture underlying the process of written and spoken word forms, the functional and neural architecture of word meaning and the neurocognition of syntactic processing. As the word architecture indicates, we are here dealing basically with structure. Other sources deal with the dynamics of brain modeling; we have briefly discussed D. J. Amit (1989), *Modeling Brain Function. The World of Attractor Neural Networks* in chapter 2. In addition we may mention A. Scott (2002), *Neuroscience – A Mathematical Primer*, and C. Eliasmith and C. H. Anderson (2005), *Neural Engineering: The Principles of Neurobiological Simulation*; see also the recent papers *Mind-reading as Control Theory*, P. Gärdenfors (2007), *Dynamics in Complex Systems*, Kurt's et al. (2009), and *The Brain, a Complex Self-organizing System*, Singer (2009). These books and articles are attempts to model brain and cognitive behavior in general. One attempt aimed directly towards language behavior, is D. Loritz (1999), *How the Brain Evolved Language*. Loritz uses systems of non-linear reaction equations to model linguistic behavior. He has some successes with phonology and certain morphological and syntactic phenomena, but is rather vague when moving from syntax to semantics. In this area there is a recent attempt by C. Eliasmith (2000), *How Neurons Mean. A Neurocomputational Theory of Representational Content*. This is noteworthy, but it is fair to say that we are only in the very early stages in our quest for understanding.

Let me for a moment retreat to simplicity and point to some early attempts to model language and brain based on neural network models. One example is the work by J. Elman on recurrent networks for grammatical discrimination; see the review of this work in P. M. Churchland (1995) and, for further references, the comprehensive survey of network models in P. S. Churchland (2002). Let me also recall a proposal within the context of optimality theory, A. Prince and P. Smolensky (1997), *Optimality: From Neural Networks to Universal Grammar*, in *Science* vol. 275, March 1997. We now have a very comprehensive exposition of optimality and harmony theory, see the two volumes P. Smolensky and G. Legendre (2006). There has been a heated debate between rule-based approaches versus network models. Stated in a very crude way the proponents of the first approach want to extend chomskian type syntactic rules into the brain, whereas the network camp believes that recurrent networks do indeed model brain in a faithful way and that language structure can be explained as an emergent behavior of such networks. The integrated connectionist/symbolic architecture of harmony theory is one attempt to bridge this gap.

I shall return to this discussion in the next section. Here I shall conclude with some further remarks and examples related to "real" brain structure. It is convenient to take the neurorecognition book by Brown and Hagoort (1999) as a starting point. They focus, in particular, on three challenges: complexity, mapping languages in the brain, and anatomical and functional variability. In addition they discuss problems connected with experimental techniques such as PET (positron emission tomography), fMRI (functional magnetic resonance imaging), and ERP (event-related brain potentials). Cognitive neuroscience is a rapidly moving field and there has been much improvement in techniques and results; see as an example the review Logothesis (2008), see also the critical discussion in Miller (2008). Not being an expert I have a limited grap of this area, but let me mention two noteworthy examples of recent progress of immediate relevance to the topics discussed in this report.

The first concerns the neuronal basis of maps of the environment and gives an example of the detailed neuronal structure of a higher level cognitive property. This is a remarkable result. From grammar and blueprints we deduce diagrams filled with labels and arrows. From neuroscience we get data and possible locations for cognitive activity. A location may sometimes correspond to a label in the blueprint, and sometimes a neural correlation may correspond to a map, but we are almost always in a "black box" situation – from locations and labels it is a long way to actual neuronal structure. This contrasts with the situation at the "top" level. We have in the earlier chapters given examples of the existence of well-defined maps between grammar and conceptual structure. It was a major advance in logic to provide a mathematical structure to the interface between syntax and semantics, and we could use these techniques to construct a well-defined map between attribute-value grammar formalisms and the category of conceptual spaces. With this amount of mathematics on board we could frame precise questions about the relationships/maps/causal connections between e. g. common nouns, natural kinds, convex domains and domains of attraction for dynamical systems. The article *Microstructure of a spatial map in the entorhinal cortex* by Hafting, Fyhn, Molden, Moser and Moser (2005) deals with a similar success story at the neuronal level. We are actually dealing with rats, but if a similar class of structures exist at the human level, they would be part of the neuronal substructure of the category of conceptual spaces. I quote from the abstract of the paper:

> The ability to find one's way depends on neural algorithms that integrate information about place, distance and direction, but the implementation of these operations in cortical microcircuits is poorly understood. Here we show that the dorsocaudal medial entorhinal cortex (dMEC) contains a directionally oriented, topographically organized neural map of the spatial environment. Its key unit is the 'grid cell', which is activated whenever the animal's position coincides with any vertex of a regular grid of equilateral triangles spanning the surface of the environment. ··· The map is anchored to external landmarks, but persists in their absence, suggesting that grid cells may be part of a generalized, path-integration-based map of the spatial environment.

Our next example deals with grammatical structure. As noted by Brown and Hagoort (1999), the majority of PET and fMRI language studies have dealt with single-word processing. But, as they emphasize, "the primary goal of speaking, listening and reading lies in the message that is to be produced and understood. Message-level production and comprehension requires the activation and real-time co-ordination of several levels of linguistic and non-linguistic information. This complexity is central to the essence of language, and presents a major challenge to cognitive neuroscientists, which has to date basically not been taken up." This was written in 1999. Since then the situation has started to change. Our second example, *Who did what to whom? The neural basis of argument hierarchies during language comprehension* by Bornkessel, Zysset, Friederici, von Cramon and Schlesewsky (2005) comes from the research group of A. D. Friederici at the Max Planck Institute for Human Cognitive and Brain Sciences. I quote from the abstract of the paper:

> The present fMRI study aimed at identifying neural correlates of the syntax-semantics interface in language comprehension. This was achieved by examining what we refer to as "argument hierarchy construction", i.e. determining which participant in a sentence is "Actor" and which is the "Undergoer" of the event expressed by the verb. In order to identify the neural bases of argument hierarchy processing, we manipulated three factors known to influence the complexity of argument hierarchy constructions in German, namely argument order, verb class and morphological ambiguity. Increased argument hierarchization demands engendered enhanced activation in a network of inferior frontal, posterior superior temporal, premotor and parietal areas. Moreover, components of this network were differentially modulated by the individual factors. In particular, the left posterior superior temporal sulcus showed an enhanced sensitivity for morphological information and the syntactic realization of the verb-based argument hierarchy, while the activation of the left inferior frontal gyrus (pars opercularis) corresponded to linearization demands and was independent of morphological information. We therefore argue that, for German, posterior superior temporal and inferior frontal regions engage in the extraction of actorhood from morpho-syntactic structure and in the sequential realization of hierarchical interpretative dependencies, respectively.

This quote is but an introduction to a vast corpus of highly important work by Friederici and her group in Leipzig. Of particular interest is the identification of two separate brain networks for "simple" versus "complex" syntactic processing; see Friederici et al. (2006), Brauer and Friederici (2007), and Bahlmann et al. (2008). It is further of relevance to note that the grammar of mathematics shows some striking differences from the grammar of natural languages, see Friedrich and Friederici (to appear).

There has been much impressive progress in brain science, but much remains to be understood. I have given a few examples of recent research. The reader may also want to consult current work on the "mind of the fly"; see the paper by Liu et al. (2006) on memory traces in the *Drosophilia brain*, the review article *Into the mind of a fly* by L. B. Vosshall (2007), and the interpretative essay by C. Desplan (2007). We understand much at the neuronal level, we understand something of how neurons form functional networks, but as yet very little of how these networks form the basis of mind, language and behavior. In the next section we shall, nevertheless, review what little we know.

3.2 Closing the gap

Blueprints. A convenient starting point for this review is the "neurocognition of language" book edited by Brown and Hagoort (1999). The surveys in this book separate into three levels. At one end we see the perspective of the linguist, which rests on a long research tradition in the classical disciplines of phonology, syntax and semantics. At the other end we have the perspective from the neuroscience research community, who by now is in command of a vast and extremely detailed knowledge of brain structures. But, as we have repeatedly emphasized, when the linguists try to reach deeper into mind and brain and the neuroscientists try to explain how higher cognitive functions emerge out of bare structure, we are in a somewhat uncertain middle ground. The two middle parts of Brown and Hagoort (1999) survey various attempts to bridge this gap. In one part the linguistic analysis is supplemented by an account of cognitive architecture, building on a rich research tradition in cognitive psychology. Here we find mid-level blueprints of the speaker, the listener and the reader. In a second part we find a survey of steps towards a neurocognitive architecture of language, aiming to connect the blueprints of cognitive psychology with basic brain structure. What is particularly attractive with the Brown and Hagoort book is the effort to spell out the interaction between the different parts. On the one side the blueprints are so constructed to be consistent with the linguistic analysis, and the other side the blueprints are intended to serve as guiding principles in the identification of the basic neural architecture.

As an example we shall take a closer look at the blueprint for the speaker, as presented in the chapter by Levelt (1999). The speaker blueprint has two main components: (i) the rhetorical/semantic/syntactic system, and (ii) the phonological/phonetic system. The two systems have access to several knowledge sources: (i) knowledge of external and internal world, (ii) a mental lexicon, and (iii) a syllabary.

We shall in a few words indicate how the system is intended to function. In the first stage, the *conceptual preparation stage*, a message is generated making use of the external/internal knowledge base. According to Levelt this message has ultimately to be a "conceptual structure", consisting of lexical concepts, that is concepts for which there are words in the language. Next stage is the *grammatical encoding*, where the lexical concepts activate the corresponding syntactic words (lemmas) in the mental lexicon. The speaker uses this lexical- syntactic information to build an appropriate syntactic form, the surface structure, which is the output of the *rhetorical/semantic/syntactic* subsystem. As soon as a lemma is selected from the mental lexicon, its associate code becomes active. And with this added input the combined *phonological/phonetic* system converts the surface structure through *morpho- phonological* and *phonetic* coding and, as a final step, *articulation* to give the final output, *overt speech*. A flow diagram of the system is found on page 87 of Levelt (1999). The reader will note several feed-back loops in this flow diagram, indicating possibilities for correcting the message at that stage.

Levelt notes the basic consensus in the research community about speech production architectures. We have only hinted at the main components in the given blueprint for the speaker, the reader is advised to consult Levelt (1999), and the references quoted there, to see the more detailed structure of, in particular, the phonological/phonetic system. Let me add a remark on the possible compatibility of this architecture with current linguistic analysis, in particular to the analysis reviewed in earlier parts of this book of the theory of conceptual spaces and the technology of attribute-value representational forms. It is straight forward to see that the theory of conceptual spaces, see the section on the structure of semantic space, can be used to give a precise model-theoretic structure to the external/internal knowledge base, and that the format of attribute-value representational forms (see the section on the structure of grammar) can be used as the initial conceptual structures in the speech production system. With this input the grammatical encoding falls within the domain of attribute-value grammars and unification processes. A system of unification grammar is actually used by Levelt, see Kempen and Vosse (1989) as well as our brief exposition of this grammar formalism in chapter 2.

The *Neurocognition* book is certainly not the only attempt to present a comprehensive view of language and cognition. Here I would also like to draw attention to the book by S. M. Kosslyn and O. Koening, *Wet Mind*, from 1992. It is interesting to compare the blueprint of the speaker in the Kosslyn-Koening book (the interested reader is referred to the diagram on p. 249 of the book) with the blueprint presented by Levelt. There is a basic or global compatibility of subsystems, but there are many differences in details. There are several reasons for this. One may be the general difficulty or ambiguity in the task of putting a "fine" structure on observational data, a problem in every inductive or empirical science. The other may be due to a difference in research focus, which, in particular, is seen in the discussion of the fine structure of the phonlogical/phonetic system. Compared to the "fine structure" of the Kosslyn-Koening blueprint the Levelt blueprint is much more "coarse grained" and it is also in its overall structure closer to current theories of grammar.

Turning from grammar and psycholinguistics to cognitive neuroscience blueprints can, as emphasized by Levelt, be seen as frames for a research programme. They summarizes in their architectural structure much of current knowledge from linguistics and cognitive psychology. And it is conjectured that an agenda for further research in cognitive neuroscience can be read from these blueprints. But before entering into details about models and algorithms we need to reflect further on issues of complexity and reducibility.

Writing equations. In the introductory section we briefly touched some issues connected to reductionism in science. We noted that traditional methodology of science often postulated a "seamless connection" between natural phenomena, even if they operate on different levels and seem to be of different kinds. This has been a belief which has been very productive in the development of modern science; we understand and hence control by reducing the complex to the simple. A powerful statement of this belief can be found in the 1980 Inaugural Lecture as Lucasian Professor at Cambridge University by Stephen Hawking; see the discussion in Fenstad (2007). Hawking starts the Inaugural with the following prediction: "In this lecture I want to discuss the possibility that the goal of theoretical physics might be achieved in the not too distant future, say, by the end of the century. By this I mean that we might have a complete, consistent and unified theory of physical interactions which would describe all possible observations." And reductionism according to Hawking goes far beyond physics proper. He almost says that if we have the final equations of physics, then everything else will follow. He notes Dirac's reduction of chemistry to physics, but is careful to explain that this is only in principle. Real systems are too complex for detailed calculations and one has "to resort to approximations and intuitive guesses of doubtful validity." Biology comes next, but he notes with some regret that "although in principle we know the equations that govern the whole of biology, we have not been able to reduce the study of human behaviour to a branch of applied mathematics".

Language and cognition are basic components of human behaviour. It is not absolutely clear from the text, but these phenomena are presumable also included in Hawking's reductionism, and hence that equations for language and mind can, at least in principle, be written in terms of brain structure. This is controversial ground and we shall try not to become too deeply ensnared. But a few remarks related to mathematical modeling are in order.

It is generally accepted that an understanding of consciousness and self-awareness is basic for any "final" theory of language and meaning; see e. g. the discussions in Tomasello (1999) and Gärdenfors (2003). Out of a vast body of literature on consciousness and understanding let me mention one text, *The Quest for Consciousness. A Neurobiological Approach*, by C. Koch (2004). The book reports on many years of research, partly in cooperation with Francis Crick, on a neurobiological theory of consciousness. The book is not a simpleminded exercise in reductionism, but an elaborate search for the *neural correlates of consciousness*. This is an account of work in progress, no final answer is given. The text is rich in reports on experiments and discussions of recent theoretical work. But as noted by J.-P. Changeux (2004a) in his review in *Nature*, there is one missing theme: "can a model of consciousness be formalized in mathematical terms?" Changeux, echoing Hawking's belief, is not in doubt, his answer is "a resounding *yes*" – equations for consciousness can and will be written; for some references to Changeux's own efforts towards this goal, see Dehaene et al. (1998), Dehaene et al. (2003), and Changeux (2004); see also the early contribution, *Neuronal Models of Cognitive Functions*, Changeux and Dehaene (1989), which we discussed in chapter 2.

Let us pause for a moment and point out that – despite Hawking – writing equations and reductionism is not the same thing. In a recent essay in *Nature, Physics, complexity and causality*, G. Ellis (2005) notes that in a typical hierarchy of complexity, each level is linked to levels below, starting, say, from basic physics, moving through chemistry and biology, to health and the behavioural sciences; see also the discussion in Scott (2002). The crucial word is "linked". Strict reductionism asserts that causation always works upwards, thus "link" has a causal meaning in the sense that each level is explained in terms of levels below – physics is all there is. But the world is more complicated than the physicists like to believe. They agree that in complex systems equations are written separately on each level, but such equations should – at least in principle – be reducible to more basic levels, just as thermodynamics is reducible to statistical physics. This may be fine when equations are linear, but in the non- linear cases we see new phenomena emerge, which are not reducible to component parts, ecological systems, the cognitive sciences and even physics offer many examples. The fundamental point here is that in complex systems we may have both upwards and downwards causality; in addition to Ellis on physics, see the specific discussion of the cognitive sciences in G. Marcus (2004), *The Birth of the Mind*. Thus we may, in principle, agree with Changeux that it is possible to "write equations for consciousness", but at their appropriate level or levels of the cognitive hierarchy, and at the same time accept that we can, in agreement with Koch, at most search for the neural correlates of consciousness, never aim for a complete reduction of consciousness to physics and biochemistry; on this issue see, however, the recent article by Schiff et al. (2007) in *Nature*, and the comments by Shadlen and Kiani (2007) in the same issue of the journal. To conclude, I am not in this paragraph arguing for a dualism between body and mind in the traditional philosophical sense; I am simply expressing the fact that causality in complex systems is more than reductionism.

Computational neuroscience is today a thriving field; see the text *Neuroscience. A Mathematical Primer*, Scott (2002), and the recent collections of review articles in *Science*, vol 314, October, 2006. In their introduction to the *Science* review P. Stern and J. Travis (2006) note that scientist increasingly use mathematical modeling and computer simulations to study and predict the behaviour of the nervous system in areas ranging from molecules to the higher brain functions. Models and experiments are, of course, essential, but they highlight the importance of simulations because of the complexity of the current experimental situation. They note that new and fruitful collaboration between modelers and experimentalists are developing:

> The results produced in simulations often lead to testable predictions and thus challenge other researchers to design new experiments or reanalyzing their data as they try to confirm or falsify the hypotheses put forward.

This, indeed, is a description of a fruitful and two-way flow of data and designs, and it testifies to the growing maturity of computational neuroscience as a research field.

The *Science* survey is divided in three parts. In the first part Herz et al. (2006) review our current understanding of the dynamics and computations of single neurons. But single neurons are part of larger systems. In the next part of the survey Destexhe and Contreras (2006) review recent advances in our understanding of more complex systems in which neurons and networks of neurons are subjected to stochastic inputs. In the final part O'Reilly (2006) reviews the development of models of higher-level cognition. The hope, consistent with our remarks above, is that a "mature" computational neuroscience based firmly on the anatomy and physiology of the human brain can help us to understand the structure and functioning of conscious awareness and human intelligence.

The editors of the *Science* survey emphasize, in particular, one suggestion in O'Reilly's review, that the prefrontal cortex represent a synthesis between analogue and digital forms of computations. I shall return to this suggestion in the concluding part of the chapter. Let me here just note that an understanding of the interaction between the digital mode of computational grammar and the analogue computations of the neuronal system is crucial for an understanding how language and meaning is "linked" to actual brain structure and that a possible synthesis most probably will be rooted in the bi-stability associated with many non-linear dynamical systems. This has been acknowledged for a long time, see e. g. the collection P. Kruse and M. Stabler (1995), *Ambiguity in Mind and Nature. Multistable Cognitive Phenomena.* I also recommend the discussion of neuronal assemblies in chapter 11 of Scott (2002). Computational neuroscience is a main part of current system biology; for a general introduction to system biology see the recent book, *The Music of life. Biology Beyond the Genome*, by D. Noble, one of the founding fathers of the field. We note that the need for a "system approach" extends far into the market place; as an example see the current review in *Nature* by A. Abbott (2008). Abbott remarks in his review article that "when genomics matured at the turn of the century, much of the industry was convinced that individual genes would emerge as new drug targets. But that reductionist bubble soon burst ... As individual genes have fallen out of favour, 'systems' – multitude of genes, proteins and other molecules interacting in an almost infinite number of ways – have come into vogue. System biology is an attempt to make sense of all these data." He continues: "At its heart, system biology is about gathering unprecedented amounts of data ... then making sense of it through mathematical models." And the final result should be systems and algorithms for designing new drugs. Thus the pharmaceutical industry is in a very "applied way" struggling to understand and control causality in complex non-linear systems. The market needs concrete results, not only "plausible stories" and grand visions. This is an example we should keep in mind when we try to understand the mechanisms of the brain responsible for the higher cognitive functions.

It is now time to be a bit more specific. In the next two sections I turn to a few examples of models of higher-level cognitive phenomena. But allow me first one further general remark on mathematical modeling. One should think that modeling is above all a mathematical *science*. But this is not so. It is better described as an *art*, and even further as an art form not to be constrained by any dogmas of purity in methods. The modeler comes with a tool-box of methods and algorithms. And it should be part of the ethics of the trade that tools should be adapted to the phenomena, not that the phenomena should be reshaped to fit the existing tools. The latter is often a sin of the travelling physicist, when he insists that the brain is nothing but a glorified spin glass system, allowing him to use his current tool-box from statistical physics, in particular, mean field theory, without modifications. Of course, one should note similarities and exploit the generality of existing mathematical methods, but one should always welcome new tools and analogies. Here is one example, taken from a recent essay in *Nature* by J. Doyle and M. Csete (2007); see also Doyle et al. (2005) and M. A. Moritz et al. (2005). I quote the introductory paragraph of the essay:

> Chaos, fractals, random graphs and power laws inspire a popular view of complexity in which behaviours that are typically unpredictable and fragile "emerge" from simple interconnections among like components. But applied to the study of highly evolved systems, this attractive simple view has led to widespread confusion. A different, more rewarding take on complexity focuses on organization, protocols and architecture, and includes the "emerging" as an extreme special case within a much richer dynamical perspective.

Let me note that the two papers in PNAS deal, respectively, with the *architecture of the internet* and the *role of natural fire regimes in the resilience of terrestrial ecosystems.* I have included this example, not as the final word on modeling and computations, but to highlight the fact that writing equations and constructing algorithms is necessarily an open-ended activity.

Models at midlevel. We now turn to some recent models of higher order cognitive functions. This is an area of inquiry which has more often lived on metaphors than accurate science. With the advent of modern computer technology, "mind as computer" became the preferred slogan of classical AI and early cognitive science. The Turing-machine model of brain is now to a large extent discredited, but the brain do compute and current language technology is still dominated by the mind-as-computer model, being firmly based on symbols, rules and computations.

The mind-as-computer metaphor was followed by another metaphor, which took cognitive functions as parallel distributed processes in networks of (abstract) neurons, and where mind was seen as an emerging property of network dynamics; for an early and very influential text see D. J. Amit (1989). The metaphor has proved to be productive; see the recent survey in P. S. Churchland (2002). But the artificial neural networks of connectionists modeling are sometimes rather far removed from the real biology of the brain. I shall return to this shortcoming, but first turn my attention to the "gap" between the rule-based approach of symbolism and the network approach of connectionism.

We have on several occasions pointed out that a theory of language has to give an account of "facts" on many levels. At one end we need a theory of grammar with equal emphasis on the syntactic and semantic components, at the other end a theory of the structure and functioning of the brain. And the two parts must be linked. *The Integrated Connectionist/Symbolic Cognitive Architecture* (ICS) proposed by Paul Smolensky and co-workers is an attempt to provide such a link. An early contribution was the article *Information processing in dynamical systems: Foundation of Harmony Theory* by Smolensky in volume 1 of McClelland and Rumelhart (1986). Further steps towards the goal appeared in *On the proper treatment of connectionism*, Smolensky (1988), and *Tensor product variable binding and the representation of symbolic structures in connectionist networks*, Smolensky (1990). The full program was outlined in the *Science* article *Optimality: From Neural Networks to Universal Grammar*, Prince and Smolensky (1997). A comprehensive treatment is now available in the two volumes, *The Harmonic Mind — from neural computations to optimality-theoretic grammar*, Smolensky and Legendre (2006). The theory is based on three general principles. The first is (Smolensky and Legendre p.65 and p.188)

P1. *Cognitive representation in ICS.*

Information is represented in the mind/brain by widely distributed activity patterns — activation vectors — that, for central aspects of higher cognition, possesses global structure describable through the discrete data structures of symbolic cognitive theory.

The symbolic structures s are defined by a collection of structural roles r_i and fillers f_i. The structure s is then a set of constituents f_i/r_i, where the filler f_i is bound to the role r_i. In a sense the structure s is an attribute-value matrix with the roles seen as attributes and fillers as values. If we have activation vectors r_i and f_i realizing the roles and fillers, the structure s is realized as the activation vector

$$s = \sum_i f_i \otimes r_i,$$

where s is the vector sum of the tensor products $f_i \otimes r_i$. A substantial part of the theory is devoted to building representations for a wide class of symbolic cognitive structures, see chapters 7 and 8 of Smolensky and Legendre (2006). But representation is only a preliminary step to processing, which is the content of the second principle (Smolensky and Legendre, p.66 and p.192)

P2. *Cognitive processing in ICS.*
   Information is processed in the mind/brain by widely distributed connection patterns — weight matrices — that, for central aspects of higher cognition, posses global structure describable through symbolic expressions for recursive functions of the type employed in symbolic cognitive theory.

The simplest kind of a connectionist network is the linear associator which transforms an input vector $i = (i_1, i_2, \ldots, i_m)$ to an output vector $o = (o_1, o_2, \ldots, o_n)$ by means of a weight matrix $W = \{W_{kl}\}$ according to the rule

$$\iota_k = \sum_l W_{kl}\, i_l.$$

Here i_l is the activity input of unit l of the input vector, W_{kl} is the weight of the connection from input unit l to output unit k, and ι_k is the combined input received by the output unit k. Thus in this core case the output o is the linear transformation of the input i defined by the weight matrix W, $o = W \cdot i$. This is the starting point for an extensive development. The next step is the construction of trees from constituents: There are two matrices W_{cons0} and W_{cons1} such that if x and y are vectors realizing trees x and y, then the activation vector realizing the composed tree s = [x,y] is

$$s = W_{cons0} \cdot x + W_{cons1} \cdot y,$$

see chapter 8 of Smolensky and Legendre.

The combined effect of the first two principles is to show how computations at the symbolic level can be realized by connectionist computations. This makes it important to understand computations at the connectionist level, which is precisely the content of the third principle (Smolensky and Legendre, p.75 and p. 218–219)

P3. *Harmony Maximization.*

   a.   In a number of cognitive domains, information processing in the mind/brain constructs an output for which the pair (input, output) is optimal: processing maximizers a connectionist well-formedness measure called Harmony. The Harmony function encapsulates knowledge as a set of conflicting soft constraints of varying strengths; the output achieves the optimal degree of simultaneous satisfaction of these constraints.

b. Among the cognitive domains falling under Harmony Maximization are central aspects of knowledge of language — grammar. In this setting, the specification of the Harmony function is called a Harmony Grammar. It defines a function that maps input to output,

output = parse(input).

Let us spell out in more details how nets compute. In our case a net represents an activation vector a, typically realizing some symbolic construct. We assume that some of the units of a represent an input vector i and others an output vector o. The computation is started by imposing a specific input vector i on the set of input units. This activates the network and information spreads from the input units to other units of a. At each unit, except the fixed input units, there is a repeatedly update of activation value in response to input from other units. When the network stabilizes and there is no further changes in values at the units of a, we can read off the output vector o as determined by the output units. The computation is determined by the update function. In the discrete case, the update equation reads

$$\Delta a_k = \Delta t \mathrm{F}(a_k, \iota_k),$$

where $\iota_k = \sum_l \mathrm{W}_{kl} a_l$ and F is some given function characteristic of the network in question. In order to state the maximization theorem we need to restrict the class of networks by imposing a few conditions on the activation function F, the connectivity W of the network, and the updating process (Smolensky and Legendre, p.217). Briefly, the activation level at a unit will increase or decrease according to whether the total input to the unit is positive or negative; the connectivity pattern as represented by W is either feed-forward (no loops) or symmetric ($\mathrm{W}_{kl} = \mathrm{W}_{lk}$); and units changes their activity one at a time (in the discrete case). Networks satisfying these conditions are called *harmonic*. The entity to be maximized is the so-called *Harmony function* and is defined as follows

$$\mathrm{H}(a) = a^{\mathrm{T}} \cdot \mathrm{W} \cdot a,$$

where a is the activation vector, a^{T} is the transposed vector, and W the weight matrix of the network. The origin of the harmony function lies in physics, in particular, in equilibrium statistical mechanics. Paul Smolensky came from physics and knows the art, anyone else would profit by a study of chapter 3, *General ideas concerning dynamics* in Amit (1989).

We are now ready for harmony maximization (see Smolensky and Legendre, p.219):

In all cognitive domains, cognitive processes are spreading activation algorithms — a core class of which are those operating in harmonic networks — that perform soft constraint satisfaction; that is, they complete input activation vectors to total activation vectors that maximizes Harmony, optimally satisfying the soft constraints embodied in the network connections and encapsulated in the Harmony function
···

It is time to turn our attention to the second part of the Harmony Maximization principle. It asserts that knowledge of language — grammar — is realized in harmonic networks (Smolensky and Legendre, p.219). The harmonic networks has a rather simple structure, but their computational complexity is equivalent to Turing machine computability in the sense that any formal language can be specified by a harmonic grammar (Smolensky and Legendre, chapter 10). It is therefore not unreasonable to expect that the formal machinery of harmonic networks will be able to cope with the complexities of natural language. This is not a statement simply to be proved, but requires a case by case modeling or reconstruction of natural language properties within the framework of harmonic grammars.

The test case chosen by Smolensky and Legendre is the phenomenon of *split intransitivity* or *unaccusativity* in French (Smolensky and Legendre, 6.2.5 and chapter 11). The phenomenon itself seems rather simple. Many intransitive verbs can be thought of as derived from more basic transitive verbs, from *John ate pasta* we get *John ate*, and from *Karen rolled the ball* we get *the ball rolled*. We get two classes — a split — of intransitive verbs, either a class with an agentive subject (eat) called unergative, or one with patient subjects (roll) called unaccusative. But real languages are not so simple. Intransitive verbs do seem to split in two cases, +/- unaccusative, but this is related to a number of syntactic and semantic conditions.

One syntactic test used in this study is Object raising (OR) with the predicate faire. The context has the format:

[singular noun phrase] est facile a faire ···

In the example, let the argument be une souris (a mouse) and let the verbs to be tested be mourir (die) and courir (run). A French speaking person would accept "une souris est facile a faire mourir", but reject "une souris est facile a faire courir". It would be nice if we could write down a list of syntactic and semantic features, which would sharply divide the class of verbs into +/- unaccusative. But there are uncertainties blurring the dividing line between the two cases, making a "traditional" grammatical account of this feature difficult. Enter Harmony Grammar

(Smolensky and Legendre, p.231):

> The semantic and syntactic tendencies characterizing the accept-
> ability of test sentences can be formulated as a set of soft constraints
> in HG. For any given sentence, the grammar assigns a Harmony value:
> positive values indicate grammaticality, negative values indicate un-
> grammatically, and values close to zero indicate a marginal level of
> grammaticality or ungrammaticality.

A full account is given in chapter 11 of Smolensky and Legendre.
The HG constructed in this chapter has among it's inputs 4 syntactic
test (among them OR), 143 verbs (including "mourir" and "courir"),
two semantic features of the argument (animacy, volitionally), and two
aspectual features of the verb (telicity and progressivizability). There
will be one output unit giving the degree of grammaticality of the input
sentence. And the grammar performed well, correctly accounting for all
but 2 of the corresponding 760 acceptability judgements, see Legendre,
Miyata, and Smolensky (1990 a, b).

Let me add a few remarks on input and harmony: ICS theory distinguishes between a symbolic level and a lower level of activation patterns. If we want an account of how grammar and other aspects of
higher cognition emerges out of brain, we would expect the patterns to
record the activity of actual brain structures. But the nets constructed
in chapter 11, LNet, HNet and HNet' are computational devices which
accept certain inputs (context, predicate, argument, hidden structure)
and produces through the computational process of harmony maximization a number giving a measure of the grammatical acceptability
of the input. Inputs to the computation will be in form of tensor product representations starting from basic filler/role constituents, e.g. in
the case of the sentence "une souris est facile a faire mourir" the vector
representation will be

$$s = OR \otimes r_{\text{contex}} + mourir \otimes r_{\text{pred}} + une\ souris \otimes r_{\text{arg}} + 2 \otimes r_{\text{struc}},$$

(where the last term derives from the hidden structure constituent).
The input vectors and the weight matrix of the network are assigned
"suitable" numerical values. The calculation of the harmony funtion,
$H(s) = a^T \cdot W \cdot a$, is vastly facilitated by the decomposition following from the HG Soft-Constraint Theorem (Smolensky and Legendre,
p.219). These are powerful and sometimes very successful techniques.
But these are techniques at mid-level, and there are still steps to take
to provide a satisfactory link between grammar and brain.

It is in this context that Optimality Theory can be properly seen.
Harmonic Grammar lives by numerical calculations, but this is not the
case in standard grammar, nouns are not numbers. But if we hold the

view that optimization is also the paradigm for (symbolic level) grammar, we should "seek a kind of well-formedness computation that is like Harmony maximization in HG, but with no numerical arithmetic (Smolensky and Legendre, p.455)". This is a basic motivation behind Optimality Theory, and chapters 12 to 19 of SL deals with Optimality Theory from this perspective at great length, but primarily at the symbolic level. In chapter 20 the connection between Optimality Theory and Harmonic Grammar is studied, but the last word on the reduction of OT to HG has not yet been said.

Remark. Even if logic is not the main topic of interest in our discussion of optimality theory, we should point out that there has in recent years been a number of studies on the relationship between optimality theory and logic, in particular, preference logic; for a comprehensive survey see Benthem (2008).

I have in chapter 2 sketched an another attempt to close the gap between meaning and brain, the missing link being geometric structure. Via a more gently executed rule-based approach we are now in a position to move from phonology and syntax to a conceptual structure as explained above, see the section on the geometry of semantic space. The conceptual part has two components, first, the representational form in the form of an attribute-value matrix (e. g. a situation schema), and, second, the model structure, where we have opted for the format of conceptual spaces. A conceptual space is a collection of domains, where each component domain is a standard model structure enriched with an intrinsic geometric structure.

And it is geometric structure which points to a link between concept and brain, see the discussion in chapter 2. We spell this out in some detail: A natural kind is, as explained above, a property related to one of the domains of a conceptual space, more specifically a natural kind is a convex region of a domain. This is the view from the language side. Seen from the brain modeling side we recognize that the various dynamic processes in the brain have associated geometric constructs. This was explicitly used by Thom in his early attempts towards a topological semantics for language, Thom (1970, 1973). Thus similar to the prototypes and convex regions in the domains of a conceptual space we have attractors and domain of attraction in the "potential surfaces" of topological semantics. If one identifies the two, and there are certainly mathematical theorems to prove in this connection, a link is established between language, concept and brain. Note, that this is an account very much consistent with the discussion in P. M. Churchland (1995) of coding and pattern recognition. We have to admit that much of this, however, is at the present time a yet to be realized research programme, but the theory of dynamics as discussed in the early models of Amit (1989) and in the control theoretic approach, see chapter 8 of Eliasmith and Anderson (2003), seems to point in the right direction. There is also some recent research on attractor dynamics and memory which can be taken to support our account, see T. J. Wills *et al* (2005). But one need to be extremely careful in choosing the "right" geometrical representations on both the conceptual side and the brain side to make sense of the connection.

Remark. In this connection let me also recall another suggestion that I made several years ago, see Fenstad (1998) and chapter 2. My remark at the time was "that while there certainly is a correspondence between cognitive function and neural structure and activity states, the correspondence is not necessarily one-to-one; different activity areas with different architectures may generate similar geometries". This point of view is consistent with some ideas in current systems biology, see Noble (2006). And some recent experiments give further support to the view that there is indeed a many-many correspondence between brain locations and higher order structures in the explanation of cognitive functions, see the work by Schlaggar, Petersen and their co-worker in PNAS, Dosenbach et al. (2007). This paper presents some evidence for a hypothesis developed first by Petersen that two networks, rather than two areas of the brain keep the adult mind concentrated on long-term achievement. If true, the many-many correspondence between brain structure and cognitive geometry discussed in this paragraph, would have consequences for anyone searching for the neural correlates of any higher cognitive function.

In our enthusiasm for (geometric) structure we should not forget algorithms. In the Harmony approach, algorithms are basic; we read already on page one of Smolensky and Legendre (2006) that "the human mind/brain is a computational system $\cdots$" There is, of course, no contradiction between the two points of view. Algebra/algorithms and geometric structure play a complementary role in mathematics and mathematical model building. It is perfectly possible to study topology/geometry and algebra as autonomous subjects, but extra power and deeper insight are added when the two are combined. You may for example start out with the "geometry" and look upon the "algebra" as a way of introducing coordinates in order enable explicit computations. This is the point of view of algebraic and differential geometry. This is also the view of the model theoretic approach to logic as presented in Barwise and Feferman (1985), where structure comes first and language, proofs and algorithms represent in a very specific sense a coordinatization useful for explicit computations; for an elaboration of this point of view to a study of logic and human reasoning, see Stenning and van Lambalgen (2007). Adapted to our situation we may introduce some semantic structure, for example in the format of a conceptual space or a system of situation semantics, as the primary object of study and look upon the attribute-value matrices (or, as an alternative, the filler/role constituents of Harmony theory) as providing us with an algebra introducing suitable coordinates for the representation of geometric/semantic structure. At this point the reader should refer back to the discussion at the end of chapter 2.

The interplay of algorithms and structure is a theme that runs through much of mathematics and logic; see Fenstad (2002). When they "properly" interact, the result is powerful. But they may also be pursued separately. With reference to language the insistence of Chomsky on the autonomy of syntax is one example of "pure" algorithmics, and the viewpoint of Thom (1970), and to some extent the cognitive grammarians, is an example of "pure" geometry. To add a personal note, it was the simultaneous reading of Chomsky (1965) and Thom (1970) that made me see the limitations of both point of views and to suggest the property/attractor linkage, see Fenstad (1978). The cognitive grammarians, in particular, should take note that Chomsky and Thom together pack more power than either Chomsky or Thom alone.

We should, perhaps, add one final remark on computations. Much has been made of "the great computational divide" – of the difference between the symbolic/linear computations of grammar and the connectionists/parallel computations of the real brain, see e. g. chapter 5 in Donald (2001). Transcending this divide was indeed one of the motivating forces behind both the "integrated connectionist/symbolic cognitive architecture" of Harmony theory and also of the control theoretic approach, see the next section. At the neurobiological level we should also recall the suggestion reported above that the prefrontal cortex in some way represents a synthesis between analogue and digital forms of computations, with the phenomenon of bistability associated with non-linear systems being part of the explanation. We have today sufficient modeling tools to deal with this "divide". There are even hard mathematical results bridging this "divide". We know today that the dynamics of neural nets can be seen as fixed-point computations in nonmonotonic logics. For an early contribution see Balkenius and Gärdenfors (1991); a recent study is Blutner (2004b), *Nonmonotonic Inferences and Neural Networks*. The mathematics is thus clear, but what is less clear, is if there is a sharp dividing line somewhere between brain and grammar concerning the "proper" use of computational strategies in the modeling process. An understanding of this would need a better understanding of the neural correlates of computations. We should add that there are also cases where a mixed computational strategy may be warranted by the linguistic facts, see e. g. the discussion of regular and irregular verbs in Pinker (1999). My general advice would be that structure should guide the choice of computational tools in the modeling task. To take one example, I have proposed to identify natural kinds and domains of attraction, but identity "in extension" is not the same as identity "in intension", the same object in different disguises may call for different computational tools – in our example we have seen how this happens when formal semantics meets dynamical system theory.

Adding time. We next need to add some remarks about time. *Time* as a variable or parameter occurs at many places in the study of language and meaning. In the analysis of a (written) sentence or a (spoken) utterance tense is an constituent of the object under study, for example in LFG tense is in the final analysis represented as an entry in an attribute-value matrix; see chapter 1. Further, any grammatical analysis or test involves computations, and computations take place in steps, i. e. the computation develops over time. One example is the test for split intransitivity in Harmony Theory discussed above, where the maximization process is a time process. But in addition to this "intrinsic" and "hidden" occurrence of time, there are other important instances where time is an "explicit" variable/parameter in the study language and communication. A survey of the many roles of time in linguistic analysis can be found in the recent collection *The Language of Time*, edited by Mani, Pustejovski, and Gaizaukas (2005). The collection covers a vast area of traditional topics such as tense, aspects and event structure. The book has also special sections on temporal reasoning and discourse analysis. For the latter topic see also the pioneering work of Kamp and Reyle (1993).

I would like to draw special attention to the sections in the book dealing with temporal reasoning and discourse analysis. We see here an extension from a structural point of view to a process perspective. This is an extension that has many analogies in mathematics and the natural sciences, one example is the transition from a probability space approach to a stochastic process framework. A recent dynamic approach to human reasoning can be found in *Human reasoning and cognitive science* by Stenning and van Lambalgen (2007). Whereas the collection

The Language of Time firmly stays at the level of linguistic theory, a noteworthy aspect of the book by Stenning and van Lambalgen is the attempt to draw upon linguistics, logic and psychology in the analysis of human reasoning. There is even a hint of how this analysis could be grounded in brain processes. This work has now been continued, see the forthcoming report, *Language, linguistics and cognition*, in the *Handbook of the Philosophy of Language*, Baggio, van Lambalgen and Hagoort (to appear).

But to build the bridge between language and brain we still need better models of neurobiological systems. At this point it is therefore appropriate to turn to another and more recent approach to cognitive science based on ideas and methods from control theory, see Eliasmith and Anderson (2003). This is not a new approach, indeed the cybernetics of Norbert Wiener was an ambitious attempt to model brain and behavior using classical control theory; see Rosenblueth, Wiener and Bigelow (1943) and Wiener (1948). But that attempt was too much of an input-output model with insufficient attention to the structure of internal processes, and it never lived up to its initial promises. With the advent of the modern computer another metaphor, the mind-as-computer, became the dominating force in early cognitive science, until challenged by the rise of connectionism in the early 1980s.

Ideas of feedback and control were never totally absent from cognitive science, one example is the Grush emulator. The idea of an emulator was introduced in trying to understand motor control. The problem is that perceptual feedbacks are too slow to effectively control and, if necessary, adjust the motor process. Hence the introduction of a kind of internal simulator running in parallel to the real process and which makes it possible to calculate and predict on a time-scale sufficiently rapid to be able to adjust the real process; see e. g. the exposition in P. S. Churchland (2002, pp 80-85), in particular, the neural network model constructed by Pouget and Sejnowski (1997). The Grush emulator is a theoretical construct. Does it have a "real" neuronal basis? For a possibility see the recent commentary *Hippocampal cells help rats think ahead* by K. Heyman (2007).

The idea of an emulator is not necessarily restricted to motor control, for some suggested extensions see Grush (2004) and Wolpert, Doya and Kawato (2003). A recent attempt to model a broad range of higher cognitive functions in a control theoretic framework is Gärdenfors (2007), *Mindreading as control theory*. Mindreading, as used by Gärdenfors, stands for the ability of humans to represent the content of the mind of others. Gärdenfors divides this competence into representing the emotions, the attentions, the intentions, and the beliefs and knowledge of others, and he uses control theory in his attempt to model these examples of higher cognitive functions in humans. Let me note in passing that this emphasis on control theory, which is a way of coming to terms with "information flow", is currently receiving much attention among biologists trying to move beyond the fact-recording state of analysis; see as an example the recent essay *Life, logic and information*, Nurse (2008).

Another recent attempt to use the ideas and techniques of modern control theory to model neurobiological systems is Eliasmith and Anderson (2003), *Neural Engineering*; see also the thesis *How Neurons Mean*, Eliasmith (2000), and the the general discussion in *Moving beyond metaphors*, Eliasmith (2003). Since our topic is grammar and brain, we shall limit ourselves to a brief description of the approach as presented in Eliasmith and Anderson (2003), with particular emphasis on similarities and differences to the Harmony Theory of Smolensky and Legendre (2006).

Both approaches divide the modeling task in three parts: representations, computations/transformations, and dynamics. We turn first to representations. In Harmony theory information is represented by distributed activation vectors, building up more complex representations through tensor products of filler/role constituents; see principle **P1** on page 76. In the control theoretic approach the representation is kept closer to the neurobiological level. Representations are encoded information, in this case the encoding procedure is "the transduction of stimuli by the system resulting in a series of neural 'action potentials' or 'spikes' (Eliasmith 2003)". The process of decoding is equally important in order to extract information about the stimulus from the spiking neurons. In general we have an expression of the form

$$a_i(x) = G_i[J_i(x)],$$

where x denotes some stimulus to the system and $a_i(x)$ is the resulting neural firing rate of some neuron a_i. In the simplest case we assume that x stand for some scalar magnitude, but the method extends to vectors, functions and even vector fields. We also have to make a distinction between population codes (as in this example) and timing codes and to their combinations, leading to an encoding into temporally patterned neural spikes over population of neurons. $J_i(x)$ represents the soma current and can be decomposed into a background current J_i^{bias} and a driving current J_i^d. In the simplest case we can assume that J_i^d is a linear function, $J_i^d = c_i x$, thus

$$J_i(x) = c_i x + J_i^{bias}.$$

The response function G_i is determined by the intrinsic properties of the neuron. Standard theory (minimizing mean square error) proposes a linear decoder of the form

$$\hat{x} = \sum_i a_i(x)\pi_i,$$

where the explicit form of π_i is determined through the minimization process.

The next step concerns computations/representations. The question is simply this, what is the representational strength of the system, what can it compute. We recall how this was handled by weight matrices in the Harmony approach; see principle **P2** on page 77. Here we extend the representational decoder to a suitable form of a transformational decoder, i.e. instead of computing some estimate of the signal x, we compute an estimate of some transformed version $f(x)$ of the signal. In this way we open up for the possibility of representing higher cognitive functions in the system. To fix ideas, take the simplest possible example, the identity transformation. In this case the purpose is to send some (scalar) signal represented in a neuron population a_i to another location represented by a population b_j. Here the transformation is $f(x) = x$. If we assume that the estimate of x as given by above formula is acceptable, we get the desired result by substituting this estimate for $y = f(x)$ in the expression for $b_j(y) = Gj[J_j(y)]$,

$$b_j(x) = G_j[\sum_i w_{ji} a_i(x) + J_j^{bias}],$$

where the weight matrix of the combined system of a_i and b_j neurons is given by $w_{ji} = c_j \pi_i$. Much space is given in Eliasmith and Anderson to determine the range of transformations representable in this way. This is in exact parallel to the theory in chapter 8 of Smolensky and Legendre. It remains to discuss the dynamics of the control approach.

In the Harmony Theory of Smolensky and Legendre (2006) the dynamics of the system was handled by a computational strategy relying on the maximization of the Harmony function; see principle **P3** on page 77. In the approach of Eliasmith and Anderson modern control theory is called to service. Control theory has a wide range of applications, and the general formulation needs to be adapted to the case under study. Here we are interested in a control theoretic description of neural populations; see section 8.1.2 of Eliasmith and Anderson (2003). Briefly, the system description can be summarized in the following diagram:

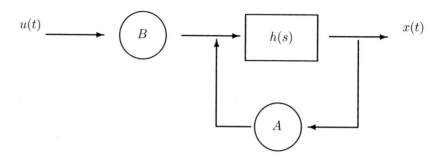

In this diagram $x(t)$ is the state variable, $u(t)$ is the input or control vector, A is the dynamics matrix, and B is the input matrix. h is the transfer function of the system; in the neurobiological applications it is derived from the synaptic dynamics of neurons; see chapter 8 of Eliasmith and Anderson (2003) for a full explanation. The dynamics is given by the following equation

$$x(t) = h(t) * [A\,x(t) + B\,u(t)],$$

where $*$ is the convolution product

$$h(t) * [A\,x(t) + B\,u(t)] = \int_0^t (A\,x(s) + B\,u(s))h(t-s)\,ds.$$

Substituting this expression for $x(t)$ in the formula for $a(x)$ given above we get the following equation for the dynamics of a neurobiological system

$$a_i(x(t)) = G_i[c_i(h_i(t) * [A\,x(t) + B\,u(t)] + J^{bias}],$$

see chapter 8 of Eliasmith and Anderson (2003), where a number of examples are presented. But this is only the starting point of a more comprehensive theory of neural dynamics. In section 8.4 of the book the authors extend their discussion to a wider class of attractor networks. In this way the discussion links up with our previous discussion of properties, natural kinds and attractors. Their last topic is statistical inference and learning, an issue we touched in the section on evolution above.

Perhaps, we should at this point add a few words on mathematics and models. In a neurobiological context the basic "construct" is often some assembly or "network" of interacting nodes or "neurons". The modeling tools you bring to the discussion of this construct depends upon where you come from. If you are "only" a cognitive scientist, you will, perhaps, remain close to the networks, the connection weights, simulations and learning; see the survey by P. S .Churchland (2002). If you are a traveling physicist you will immediately call upon statistical mechanics and mean field theory; see Amit (1989). A statistician will identify the dynamics of the structures as certain kinds of Markov processes and call for the use of statistical decision theory; see Ripley (1996). A mathematician will refer to the well developed theory of dynamical systems; see Hoppensteadt and Izhikevich (1997), and the engineer will describe the system in control theoretic terms. There are good points - but also non-negligible limitations - in every approach, and the perfect modeler should be well aware of all of them. But time is limited and no one is perfect, so we have to live with a plurality of partial perspectives and models of the same observables, whether they be objects or behaviors. But this can be turned into an advantage. An excellent strategy to enhance our understanding is to compare models, e.g. to compare computations at network level and computations at higher levels. This is one example, but it is a general fact that observable phenomena often can be modeled on different levels, a classical example from physics is thermodynamics and statistical mechanics. But whereas a plurality of models is good for insight, we have to retreat to simplifications and standardizations when we come to technology and applications, current language technology is a good example.

A final word on metaphors. The brain is not a computer, nor a neural network or a control system. Eliasmith (2003) makes the following remark in his discussion of representations and computations:

> . . . it is important to realize that this way of characterizing representation and computation does not demand that there are "little decoders" inside the head. That is, this view does not entail that the system itself need to decode the representations it employs. In fact, according to this account, there are no directly observable counterparts to the representational or transformational decoders. Rather, they are embedded in the synaptic weights between neighboring neurons. That is, coupling weights of neighboring neurons indirectly reflect a particular population decoder, but they are not identical to the population decoder.

This is an important point well known in the modeling of physical systems, e.g. the use of Hilbert space theory is indispensable in the modeling of hydrodynamical systems, but there is no infinite Hilbert space seen in actual ocean waves. And it is exactly an understanding of this point which allows free access to the mathematician's tool box in the modeling task in both the natural and human sciences.

3.3 What little we know

This is not meant to express a pessimistic attitude towards further advances in the understanding of language and brain. As has been famously said (see the reference in Fenstad, 2007), science is shaped by ignorance – by what we do not know. Thus these lines should on the contrary be read as an expression of optimism, and meant to point to further possibilities and new challenges, well within the range of what is do-able.

Science consists in doing what is do-able. This is in most cases a good strategy, but sometimes it may be permissible to speculate a bit on, or even beyond, the border of what can currently be done. This is a sin we have already committed, see the discussion on page 85 of the Grush emulator and its possible neuronal basis. Here, I shall add some remarks and suggestions on a very traditional language technology activity, the construction of a question/answering system with some visual inputs relating to color and shapes. Almost all research in this area rely on the "model = data base" assumption, where the data base part need neither have any direct links to actual brain structure nor to how the brain works. The system imitates, but does not model how brain, vision and language actually function. There are now some hard bits of science and technology which encourage us to move beyond the limits set by current language technology, indeed, at mid-level we have both an "algebra" and a "geometry" which can now be exploited in the modeling task. To fix ideas let us once more return to an early example of language technology, a question-answering system developed by E. Vestre (1987); see the exposition in Fenstad et al. (1992). The basic architecture follows a familiar pattern:

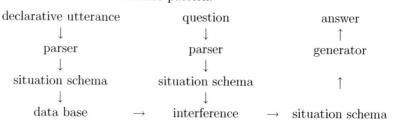

This is a technology application based on the simple assumption that "model = data base". The system can be updated by new facts as indicated by the left column. Sentences are represented by situation schemata, and a special algorithm was developed to extract basic facts from the schemata and to add them to the data base. A question is asked resulting in an incomplete schema. This schema acts as a query to the data base and produces an answer in form of a complete schema, which in turn generates the appropriate response. We note that unification and logic programming are the technical tools used in the analysis.

If we are allowed a brief moment of speculation, we could argue that at a very general level this architecture can also be used in a broader cognitive context. The right column will then have to be modified to represent a "blueprint" for a speaker or, more generally, an actor. The left column must in a similar way be modified to "blueprints" for the reader or the listener. And we must in addition make allowance for visual and other types of perceptual inputs, see Forsyth and Ponce (2003) and Ma et al. (2006) for basic technologies concerning vision and Grenander and Miller (2007) and Hallinan et al. (1999) for the broader area of pattern analysis. The middle column will represent some "attention mechanism" and will determine the appropriate context and form of response. The first move would be to replace the data bases of simple language technology with the appropriate "geometry", e.g. the category of conceptual spaces, see Gärdenfors (2000), with an added perspective from situation semantics. This would include a situational perspective, a possible partiality of models, and an independent analysis of properties. Note that a situational perspective and partiality of models in some sense function as an attention mechanism, thus possibly reducing ambiguity and uncertainty in the interpretation process. Correspondingly, the situation schemata need to be enriched to a suitable "algebra", e.g. in the form of an extended attribute-value matrix algebra. And finally, the techniques based on unification and logic programming will have to be extended to handle more complex "interference" situations, see Stenning and van Lambalgen (2007). This simple architecture has a rather static structure corresponding to a single question-answer setting and needs to be suitably enriched to account for the dynamics of reasoning and discourse, see Stenning and van Lambalgen (2007) and Baggio, van Lambalgen and Hagoort (to appear) for some recent and promising work in this direction.

So far we are within the domain of what is do-able. We remind the reader that we are still at the phenomenological level of the theory of mind as discussed in chapter 2. This level is important as a platform for language technology, but a next step is necessary for the science. But when we try to take the next step and move from the geometric model space and ask how it is grounded in actual brain structure, we are crossing the border line of what is today do-able. Concerning vision there is much to be learned from current science and technology, see e. g. Rolls and Deco (2002) and Forsyth and Ponce (2003). But the real challenge would be to see how the geometric account of color and shape can be accounted for in a "true" neurobiological framework; see Eliasmith and Anderson (2003) for a hint of how to proceed. Similarly we need to see how any proposed "algebra" at midlevel is related to the brain networks of syntactic processing identified in the work of Friederici and her group in Leipzig; see Bornkessel et al. (2005), Friederici et al. (2006), and Brauer and Friederici (2007). And the real challenge is then to understand how the midlevel account of the syntactic-semantic interface, as discussed in earlier part of this chapter, is related to these syntactic and semantic brain networks. Being a mathematician I would like to establish a suitable "commutative diagram" connecting the levels. This will be no easy task. But a promising first step has recently been published by Sahin et al. (2009). They show that different kinds of linguistic information are sequentially processed within Broca's area. In a commentary Hagoort and Levelt (1999) argue that in the particular case of speech production this neuronal process is compatible with the computational or midlevel blueprint of speech discussed in Levelt (1999); see the section Blueprints on page 69.

We have seen impressive advances in our knowledge of language and brain, but the reader needs, however, to be warned. The real biology of brain activity, whether in humans or in flies, is only beginning to be understood; see the references above to recent research on memory traces in the *Drosophilia* brain; Liu et al. (2006), Vosshall (2007). And the modeling of cognitive functions is hardly a simpler task than mastering the challenges involved in drug design, see Abbott (2008). Thus the would-be conqueror must know what hard facts there are, possess as sharp tools as possible, and be thoroughly familiar with the methodological and philosophical issues involved – understanding language and meaning is, perhaps, a bit more subtle than writing a few equations.

References

NOTE: Numbers in bold face refer to the page where the work is discussed.

A. Abbott (2008), Pharmaceutical futures: A fiendish puzzle, *Nature*, vol 455, pp 1164–1167.
74, 92

K. Ajdukiewicz (1967), Syntactic Connection, in S. MacCall (editor), *Polish Logic 1920 -1939*, Oxford University Press, Oxford, UK. (This is a translation of the original article Die syntaktische Konnexität from 1935)
2, 12

D. J. Amit (1989), *Modeling Brain Function*, Cambridge University Press.
48, 50, 65, 75, 78, 82, 89

J. A. Anderson (1995), *An Introduction to Neural Networks*, MIT Press, Cambridge, Mass.
50

G. Baggio, M. van Lambalgen and P. Hagoort, (to appear), Language, linguistics and cognition, in Kempson and Fernando, editors, *Handbook of the Philosophy of Language*.
85, 91

J. Bahlmann, R. I. Schubotz and A. D. Friederici (2008), Hierarchical artificial grammar processing engages Broca's area, *Neuroimage*, vol 42, pp 525–534.
68

C. Balkenius and P. Gärdenfors (1991), Nonmonotonic inferences in neural networks, in J. A. Allen, R. Fikes and E. Sandvall, *Principles of*

knowledge representation and reasoning, Morgan Kaufmann, San Mateo, CA.
83

Y. Bar-Hillel (1964), *Language and Information*, Addison-Wesley Publ. Comp., Mass.
13, 31

Y. Bar-Hillel and R. Carnap (1953), Semantic Information, *British Journal of the Philosophy of Science*, vol 4, pp 147–157.
33

J. Barwise and R. Cooper (1981), Generalized quantifiers and natural languages, *Language and Philosophy*, vol 4.
18, 20

J. Barwise and J. Etchemendy (1991), Visual Information and Valid Reasoning, in W. Zimmermann and S. Cunningham, editorss, *Visualization in Mathematics*, Math. Ass. America, Washington DC.
42

J. Barwise and S. Feferman, editors (1985), *Model-Theoretic Logics*, Springer Verlag.
82

J. Barwise and J. Perry (1983), *Situations and Attitudes*, The MIT Press.
20, 37

J. van Benthem (2008), For Better or Worse: Dynamic Logics of Preference, in S. O. Hanson and T. Gruene-Yanof, editors, *Preference Change*, Springer.
81

J. van Benthem and A. ter Meulen, editors, (1997), *Handbook of Logic and Language*, North-Holland.
12, 20

L. Bloomfield (1955), Linguistic Aspects of Science, in O. Neurath et al. (1955), pp 219–277.
9

R. Blutner (2002), Lexical Semantics and Pragmatics, in F. Hammel et al., editors, *Linguistische Berichte*, vol 10.
32

R. Blutner (2004a), Pragmatics and the Lexicon, in L. R. Horn and G. Ward, editors, *Handbook of Pragmatics*, Blackwell, Oxford
33

R. Blutner (2004b), Nonmonotonic Inference and Neural Networks, *Synthese*, vol 141(2).
83

I. Bornkessel, S. Zysset, A. D. Friederici, D. Y. von Cramon and M. Schlesewsky (2005), Who did what to whom? The neural basis of argument hierarchies during language comprehension, *Neuro-Image*, 26 (2005), pp 221-233.
67, 92

J. Brauer and A. D. Friederici (2007), Functional neural networks of semantic and syntactic processes in the developing brain, *Journal of cognitive Neuroscience*, vol 19:10, pp 1609–1623.
68, 92

J. Bresnan, editor, (1982), *The Mental Representaion of Grammatical Relations*, The MIT Press.
21, 29

J. Bresnan (2001), *Lexical-Functional Syntax*, Blackwell.
21, 29

J. Bresnan and R. Kaplan (1982), Lexical -Functional Grammar: A Formal System for Grammatical Representations, in Bresnan (1982).
21

C. M. Brown and P. Hagoort, editors, (1999), *The Neurocognition of Language*, Oxford Univ. Press.
65, 66, 67, 69

G. L. Bursill-Hall, (editor) (1972), *Grammatica Speculativa of Thomas of Erfurt*, London: Longman
8

R. Carnap (1937), *Logical Syntax of Language*, Routledge and Kegan Paul, London, UK. (This is a translation of the German original *Logische Syntax der Sprache* from 1934)
12, 14

B. Carpenter (1992), *The Logic of Typed Feature Structures*, Cambridge University Press.
22

B. Carpenter (1997), *Type Logical Semantics*, The MIT Press.
20

J. P. Changeux (2004), *The Physiology of Truth*, Harvard University Press.
72

J. P. Changeux (2004a), Clarifying consciousness, *Nature*, vol 428, pp 603–604.
72

J. P. Changeux and S. Dehaene (1989), Neuronal Models of Cognitive Functions, *Cognition*, 33.
48, 72

N. Chomsky (1957), *Syntactic Structures*, Mouton and Co, The Hague, Netherlands.
2, 11, 16

N. Chomsky (1965), *Aspects of the Theory of Syntax*, The MIT Press.
2, 11, 21, 22, 29, 83

N. Chomsky (1966), *Cartesian Linguistics*, Harper and Row, New York.
9

M. H. Christiansen and S. Kirby (2003), editors, *Language Evolution*, Oxford University Press.
58

P. M. Churchland (1995), *The Engine of Reason, the Seat of the Soul*, The MIT Press.
66, 81, 89

P. S. Churchland (2002), *Brain-Wise*, The MIT Press.
51, 66, 75, 85

P. S. Churchland and T. J. Sejnowski (1992), *The Computational Brain*, MIT Press, Cambridge, Mass.
51

E. Colban (1987), Prepositional phrases in situation schemata, in Fenstad et al. (1987).
43

J. Colombo, P. McCardle and L. Freund (2009), *Infant Pathways to Language*, Psychology Press, Taylor and Francis, New York.
58

A. Copestake et al. (1995), Translation using minimal recursion semantics, in *Proceedings of the 6th International Conference on Theoretical and Methodological Issues in Machine Translation*, Leuven, Belgium.
32

P. W. Culicover and R. Jackendoff (2005), *Simpler Syntax*, Oxford University Press.
22, 29

H. B. Curry (1961), Some Logical Aspects of Grammatical Structure, in R. Jakobson (editor), *The Structure of Language and its Mathematical Aspects*, Amer. Math. Socoety, Providence, RI, USA.
2, 13, 15, 18, 20

M. Dalrymple, M. Kanazawa, S. Mchombo and S. Peters (1994), What Do Reciprocals Mean?, in *Proceedings of the Fourth Semantics and Linguistics Theory Conference*, Cornell University Working papers in Linguistics.
43

S. Dehaene, C. Kerszberg and J. P. Changeux (1998), A neuronal model of a global workspace in effortful cognitive tasks, *PNAS*, vol 95, pp 14529–14535.
72

S. Dehaene, C. Sergent and J. P. Changeux (2003), A neuronal network model linking subjective reports and objective physiological data during conscious perception, *PNAS*, vol 1000, pp 8520–8525.
72

G. Dehaene-Lambertz, L. Hertz-Pannier, J. Dubois and S. Dehaene (2008), How does early brain organization promote language acquisition in humans?, *European Review* , vol 16, pp 399–411.
58

C. Desplan (2007) Time to pick the fly's brain, *Nature*, vol 450, p 173.
68

A. Destexhe and D. Contreras (2006), Neuronal Computations with Stochastic Network States, *Science*, vol 314, pp 85–90..
73

M. Donald (1990), *Origin of the Modern Mind: Three Stages in the Evolution of Culture and Cognition*, Harvard University Press, Cambridge, Mass.
39, 47, 51, 64

M. Donald et al. (1993), Precis and Discussion of Origin of the Modern Mind, *Behavioral and Brain Sciences*, 16.
39, 47

M. Donald (2001), *A Mind So Rare*, W. W. Norton & Company.
67, 83

N. U. F. Dosenbach, D. A. Fair, F. M. Miezin, A. L. Cohen, K. K. Wenger, R. A. T. Dosenbach, M. D. Fox, A. Z. Snyder, J. L. Vincent, M. E. Reichle,

B. L. Schlaggar and S. E. Petersen (2007), Distinct brain networks for adaptive and stable task control in humans, *Science*, vol 104, pp 11073–11078.
82

J. Doyle, D. L. Alderson, L. Li, S. Low, M. Roughan, S. Shalunov, R. Tanaka and W. Willinger (2005), The "robust yet fragile" nature of the Internet, *PNAS*, vol 102, pp 14497–14502.
75

J. Doyle and M. Csete (2007), Rules of engagement, *Nature*, vol 446, p 860.
75

H. Dyvik (1993), *Exploiting Structural Similarities in Machine Translation*, University of Bergen.
23, 32, 40

C. Eliasmith (2000), *How Neurons Mean*, Washington University.
65, 86

C. Eliasmith (2003), Moving beyond metaphors, *Journal of Philosophy*, pp 493–520.
86, 89

C. Eliasmith and C. H. Anderson (2003), *Neural Engineering*, The MIT Press.
65, 82, 85, 86, 87, 88, 92

G. Ellis (2005), Physics, complexity and causality, *Nature*, vol 435 p 743.
72

J. E. Fenstad (1978), Models for Natural Languages, in Hintikka et al., *Essays on Mathematical and Philosophical Logic*, D. Reidel Publ. Company.
3, 18, 19, 20, 43, 52, 83

J. E. Fenstad (1988), Logic and Natural Language Systems, in H. D. Ebbinghaus et al., editors, *Logic Colloquium '87*, North-holland, Amsterdam, pp 27–39.
24

J. E. Fenstad (1996), Remarks on the Science and Technology of Language, *European Review*, vol 4, pp 107–120.
2

J. E. Fenstad (1996a), Partiality, in J. van Benthem and A. ter Meulen, editors, *Handbook of Logic and Linguistics*, North-Holland, Amsterdam.
12, 37

J. E. Fenstad (1998), Formal Semantics, Geometry, and Mind, in X. Arrazola et al., editors, *Discourse, Interaction and Communication* Kluwer, Amsterdam, pp 85–103.
3, 82

J. E. Fenstad (2002), Computability theory: structures or algorithms, in W. Sieg, R. Sommer and C. Talcott, editors, *Reflections on the Foundation of Mathematics*, ASL.
83

J. E. Fenstad (2004), Tarski, truth and natural language, *Ann. Pure and Applied Logic*, vol 126, pp 15–26.
2

J. E. Fenstad (2007), Changes of the Knowledge System and their Implication for the Formative Stages of Scholars, *European Review*, vol 15, pp 187–197.
71, 90

J. E. Fenstad, P. K. Halvorsen, T. Langholm and J. van Benthem (1985), *Equations, Schemata and Situations*, CSLI, Stanford.
viii, 21, 22, 29

J. E. Fenstad, P. K. Halvorsen, T. Langholm and J. van Benthem (1987), *Situations, Language and Logic*, D. Reidel Publ. Company.
viii, 21, 22, 24, 28, 29, 30

J. E. Fenstad, T. Langholm and E. Vestre (1992), Representations and interpretations, in M. Rosner et al., *Computational Linguistics and Formal Semantics*, Cambridge University Press.
22, 23, 38, 90

R. Ferrer i Cancho, O. Riordan and B. Bollobas (2005), The consequences of Zipf's law for syntax and symbolic reference, *Proc. R. Soc. B*, vol 272, pp 561–565.
63

J. Flum and M. Ziegler (1980), *Topological Model Theory*, Springer Lecture Notes in Mathematics, Springer-Verlag, Heidelberg.
42

D. A. Forsyth and J. Ponce (2003), *Computer Vision*, Prentice Hall.
91, 92

A. D. Friederici, J. Bahlmann, S. Heim, R. I. Schubotz and A. Anwande (2006), The brain differetiates human and non-human grammars: Functional localization and structural connectivity, *PNAS*, vol 103, pp 2458–2463.
68, 92

R. Friedrich and A. D. Friederici (to appear), Mathematical Logic in the Human Brain: Syntax.
68

M. S. Gazzaniga, editor, (2004), *The Cognitive Neurosciences III*, The MIT Press.
65

J. Goldsmith (2005), *Language* vol 81, pp 719–736.
10

J. Goldsmith (2007), Towards a new empiricism, in *Recherches linguistiques a Vincennes*, vol 36.
59

U. Grenander and M. Miller (2007), *Pattern Theory*, Oxford University Press.
53, 91

R. Grush (2004), The emulator theory of representation: motor control, imagery and perception, *Behavioral and Brain Science* vol 27, pp 377–442.
85

P. Gärdenfors (1991), Framework for Properties, in L. Haaparanta et al., editors, *Language, Knowledge and Intentionality*, Acta Philosophica Fennica, 49, Helsinki.
43, 45

P. Gärdenfors (1993), Conceptual Spaces as a Basis for Cognitive Semantics, Department of Philosophy, Lund University, Lund.
43, 46, 47

P. Gärdenfors (1994), Three Levels of Inductive Inference, in D. Prawitz et al., editors, *Logic, Methodology and Philosophy of Science* IX, North Holland, Amsterdam.
43, 45

P. Gärdenfors (2000), *Conceptual Spaces*, The MIT Press.
3, 23, 30, 43, 44, 45, 53, 91

P. Gärdenfors (2003), *How Homo Became Sapiens*, Oxford University Press.
65, 72

P. Gärdenfors (2007), Mind-reading as Control Theory, *European Review*, vol 15, pp 223–240.
65, 85

C. Habel (1990), Propositional and Depictorial Representation of Spatial Knowledge: The Case of Path-Concepts, in R. Studer, editor, *Natural Language and Logic*, Lecture Notes in Artificial Intelligence, Springer-Verlag, Heidelberg.
42

T. Hafting, M. Fyhn, S. Molden, M. B. Moser and E. Moser (2005), Microstructure of a spatial map in the entorhinal cortex, *Nature* vol 436, pp 801–806.
67

P. Hagoort and W. J. M. Levelt (2009), The Speaking Brain, *Science* vol 326, pp 372–373.
92

P. L. Hallinan, G. G. Gordon, A. L. Yuille, P. Giblin, and D. Mumford (1999), *Two- and Three-dimensional Patterns of the Face*, A. K. Peters.
53, 91

E. M. Hammer (1995), *Logic and Visual Information*, CSLI, Stanford.
42

Z. Harris (1951), *Methods in Structural Linguistics*, The University of Chicago Press, Chicago, IL, USA.
10, 13, 16

J. L. van Hemmen and T. Sejnowski (2006), editors, *23 Problems in Systems Neuroscience*, Oxford University Press.
65

A. V. M. Herz, T. Gollisch, C. K. Machens (2006), Modeling Single-Neuron Dynamics and Computations, *Science* vol 314, pp 80–85.
73

K. Heyman (2007), Hippocampal cells help the rats think ahead, *Science*, vol 318, p 900.
85

F. C. Hoppensteadt and E. M. Izhikevich (1997), *Weakly Connected Neural Networks*, Springer.
89

H. Howard (2004), *Neuromimetic Semantics*, Elsevier.
53

J. R. Hurford (2007), *The Origin of Meaning*, Oxford University Press.
58

R. Jackendoff (2002), *Foundation of Language*, Oxford University Press.
22, 29, 30, 58, 60

O. Jespersen (1969), *Analytic Syntax*, Holt, Reinhart and Winston, New York.
15

O. Jespersen (1975), *The Philosophy of Grammar*, Allen and Unwin, London.
15

P. J. Johnson-Laird (1983), *Mental Models*, Cambridge University Press, Cambridge.
42

D. Jurafsky and J. H. Martin (2000), *Speech and Language Processing*, Prentice Hall.
32

H. Kamp and U. Reyle (2003), *From Discourse to Logic*, Kluwer.
21, 84

M. Kay (1979), Functional Grammar, in *Proceedings of the Fifth Annual Meeting of the Berkeley Linguistic Society*, University of California.
22

M. Kay (1992), Unification, in M. Rosner and R. Johnson, editors, *Computational Linguistics and Formal Semantics*, Cambridge University Press.
22

B. Keller (1993), *Feature Logics, Infinitary Descriptions and Grammar*, CSLI, Stanford.
22

G. Kempen and T. Vosse (1989), Incremental Syntactic Tree Formation in Human Sentence Processing, *Connection Science*, vol 1, pp 273–290.
53, 54, 55, 70

C. Koch (2004), *The Quest for Consciousness*, Roberts and Company Publishers.
72

S. M. Kosslyn and O. Koening (1992), *Wet Mind*, The Free Press, New York.
61, 70

J. Koster (1989), How Natural is Natural Language, in J. E. Fenstad, I. T. Frolov and R. Hilpinen (editors), *Studies in Logic* vol 126, North Holland, Amsterdam, Netherlands, pp 591–606.
17, 18

P. Kruse and M. Stadler (1995), editors, *Ambiguity in Mind and Nature*, Springer.
74

J. Kurths, D. Maraun, C. S. Zhou, G. Zamora-Lopez and Y. Zou (2009), Dynamics in Complex Systems, *European Review*, vol 17, pp 357–370.
49, 65

G. Lakoff (1987), *Women, Fire and Dangerous Things*, Chicago University Press, Chicago, Ill.
43, 46

R. W. Langacker (1987 and 1991), *Foundation of Cognitive Grammar*, vols 1 and 2, Stanford University Press, Stanford, CA.
23, 43, 46

G. Legendre, Y. Miyata and P. Smolensky (1990a), Can connectionism contribute to syntax? Harmonic Grammar, with an application, *Proceedings of the Chicago Linguistic Society*, p 26.
80

G. Legendre, Y. Miyata and P. Smolensky (1990b), Harmonic Grammar — a formal multi-level connectionist theory of linguistic wellformedness: An application, *Proceedings of the Cognitive Science Society*, p 12.
80

A. Lenci and G. Sandu (2005), Logic and Linguistics in the Twentieth Century, in L. Haaparanta, editor, *The Development of Modern Logic*, Oxford University Press.
21

S. Lesniewski (1929), Grundzüge eines neuen Systems der Grundlagen der Mathematik, *Fund. Math.* vol 14, pp 1–81.
12

J. M. Levelt (1999), Producing spoken language: A blueprint of the speaker, in Brown and Hagoort (1999), pp 83–122.
69, 70

G. Liu et al. (2006), Distinct memory traces for two visual features in the *Drosophilia* brain, *Nature*, vol 439, pp 551–556.
68, 92

N. K. Logothesis (2008), What we can do and what we cannot do with fMRI, *Nature* vol 453, pp 869–878.
66

D. Loritz (1999), *How the Brain Evolved Language*, Oxford University Press.
65

J. T. Lønning (1987), Mass Terms and Quantification, *Language and Philosophy*, 10.
37

J. T. Lønning (1989), *Some Aspects of the Logic of Plural Noun Phrases*, COSMOS-Report no 11, University of Oslo.
37

Y. Ma, S. Scatto, J. Kosecha and S. Sastry (2006), *An Invitation to 3-D Vision: From Images to Geometric Models*, Springer.
91

B. Maegaard, J. E. Fenstad, L. Ahrenberg, K. Kvale, K. Mühlenbock and B. E. Heid (2006), KUNSTI - Knowledge Generation for Norwegian Language Technology, in *Proceedings of the International Conference on Language Resources and Evaluations*, Geneva.
32

I. Mani, J. Pustejovsky and R. Gaizauskas (2005), editors, *The Language of Time*, Oxford University Press.
84

G. Marcus (2004), *The Birth of the Mind*, Basic Books, New York.
65, 72

G. Marcus and H. Rabagliati (2009), Language Acquisition, Domain Specificity, and Decent with Modification, in Colombo et al. (2009), pp 267–285.
64

H. Meinhardt (1995), *The Algorithmic Beauty of Sea Shells*, Springer-Verlag, Heidelberg
52

A. N. Meltzoff, P. K. Kuhl, J. Movellan and T. J.Sejnovski (2009), Foundations for a New Science of Learning, *Science* vol 325, pp 284–288.
58

C. Mervis and E. Rosch (1981), Categorization of Natural Objects, *Annual Review of Psychology*, 32
46

G. Miller (2008), Growing Pains for fMRI, *Science* vol 320, pp 1412–1414.
66

R. Montague (1974), The Proper Treatment of Quantification in Ordinary English, in R. Thomason (1974), pp 247–270.
2, 16, 18

M. A. Moritz, M. E. Morais, L. A. Summerell, J. M. Carlson and J. Doyle (2005), Wildfires, complexity, and highly optimized tolerance, *PNAS* vol 102, pp 17912–17917.
75

C. W. Morris (1955), Foundation of the Theory of Signs, in O. Neurath et al., pp 77–137.
12

D. Mumford (1992), Pattern theory: a unifying perspective, in *Proc. 1st Eur. Cong. of Math.*, Birkhäuser.
53

G. L. Murphy (2004), *The Big Book of Concepts*, MIT Press.
46

O. Neurath (1955), Encyclopedia and Unified Science, in O. Neurath et al., pp 1–27.
9

O. Neurath, R. Carnap and C. Morris (editors) (1955), *International Encyclopedia of Unified Science* vol I, The University of Chicago Press, Chicago, IL, USA.
9, 14

F. J. Newmeyer (1980), *Linguistic Theory in America*, New York: Academic Press.
11

D. Noble (2006), *The Music of Life: Biology Beyond the Genome*, Oxford University Press.
74, 82

M. A. Nowak (2006), *Evolutionary Dynamics*, Harvard University Press.
59

M. A. Nowak, N. L. Komarova and P. Niyogi (2001), Evolution of Universal Grammar, *Science* vol 291, pp 114–118.
59

M. A. Nowak, J. B. Plotkin and V. A. A. Jansen (2000), The evolution of syntactic communication, *Nature* vol 404, pp 495–498.
61, 63, 64

P. Nurse (2008), Life, logic and information, *Nature*, vol 454, pp 424–426.
86

S. Oepen et al. (2004), Som å kapp-ete med trollet? Towards MRS-based Norwegian-English Machine Translation, in *Proceedings of the 10th International Conference on Theoretical and Methodoligical Issues on Machine Translation*, Baltimore, MD.
32

S. Oepen et al. (2007), Towards Hybrid Quality-Oriented Machine Translation; On Linguistics and Probabilities in MT, in *Proceedings of the 11th International Conference on Theoretical and Methodoligical Issues in Machine Translation*, Skövde, Sweden.
32

A. Okabe, B. Boots and K. Sugihara (1992), *Spatial Tessellations, Concepts and Applications of Voronoi Diagrams*, J. Wiley, New York, NY.
46

R. C. O'Reilly (2006), Biologically Based Computational Models of High-Level Cognition, *Science*, vol 314, pp 91–94.
73

G. A. Padley (1976), *Grammatical Theory in Europe 1500 -1700. The Latin Tradition*, Cambridge: Cambridge university Press.
9

J. Peck (1987), editor, *The Chomsky Reader*, Pantheon books, New York.
16

H. Pedersen (1972), *The Discovery of Language. Linguistic Science in the 19th Century*, Indiana University Press.
6

S. Peters and D. Westerståhl (2006), *Quantifiers in Language and Logic*, Clarendon Press, Oxford.
20

J. Petitot (1995), Morphodynamics and attractor syntax, in Port and van Gelder, editors, *Minds as Motion*, The MIT Press.
52, 53

S. Pinker (1994), *The Language Instinct*, Harper-Collins, New York.
59

S. Pinker (1999), *Words and Rules*, Weidenfeld and Nicolson, London.
84

C. Pollard and I. Sag (1987), *Information-Based Syntax and Semantics*, CSLI, Stanford.
21, 29

C. Pollard and I. Sag (1994), *Head-Driven Phrase Structure Grammar*, Univ. Chicago Press.
29

K. Popper (1972), *Objective Knowledge*, Clarendon Press, Oxford.
40

P. Porter and B. H. Partee (2002), editors, *Formal Semantics: The Essential Readings*, Blackwell.
20

M. I. Posner and M. E. Raichle (1994), *Images of Mind*, Scientific American Library, New York, NY.
51

A. Pouget and T. J. Sejnowski (1997), Spatial transformations in the parietal cortex using basis functions, *Journal of Cognitive Neuroscience*, vol 9, pp 222–237.
85

A. Prince and P. Smolensky (1997), Optimality: From Neural Networks to Universal Grammar, *Science*, vol 275.
66, 76

J. Pustejovski (1995), *The Generative Lexicon*, The MIT Press.
32

H. Reichenbach (1947), *Elements of Symbolic Logic*, Macmillian, London, UK.
2, 13

B. D. Ripley (1996), *Pattern Recognition and Neural Networks*, Cambridge University Press.
89

E. T. Rolls and G. Deco (2002), *Computational Neuroscience of Vision*, Oxford University Press.
92

E. Rosch (1978), Prototype Classification and Logical Classification, in E. Scholnik, editor, *New Trends in Cognitive Representation*, Lawrence Erlbaum Ass., Hillsdale, NJ.
43, 45, 46

A. Rosenblueth, N. Wiener and J. Bigelow (1943), Behavior, purpose and teleology, *Philosophy of Science*, vol 10, pp 18–24.
85

W. C. Rounds (1997), Feature Logics, in van Benthem and ter Meulen (1997).
22

I. Sag and T. Wasow (1999), *Syntactic Theory: A Formal Introduction*, CSLI, Stanford.
21

N. T. Sahin, S. Pinker, S. S. Cash, D. Schomer and E. Halgren (2009), Sequential Processing of Lexical, Grammatical, and Phonological Information Within Broca's Area, *Science* vol 326, pp 445–449.
92

N. D. Schiff, J. T. Giacino, K. Kalmar. J. D. Victor, K. Baker, M. Gerber, B. Fritz, B. Eisenberg, J. O'Connor, E. J. Kobylarz, S. Farris, A. Machado, C. McCagg, F. Plum, J. J. Fins and A. R. Rezai (2007), Behavioural improvements with thalamic stimulation after severe traumatic brain injury, *Nature*, vol 448, pp 600–604.
73

P. T. Schoenemann (1999), Syntax as an Emergent Characteristics of the Evolution of Semantic Complexity, *Minds and Machines* vol 9, pp 309–346.
59, 60

A. Scott (2002), *Neuroscience – A Mathematical Primer*, Springer-Verlag, New York.
65, 72, 73

M. N. Shadlen and R. Kiani (2007), An awakening, *Nature*, vol 448, pp 539–540.
73

W. Singer (2009), The Brain, a Complex Self-organizing System, *European Review*, vol 17, pp 321–329.
55, 65

K. Skrindo (2001), *Leksikalsk Semantikk* (in Norwegian), University of Oslo.
33

C. A. Skarda and W. J. Freeman (1987), How Brain Make Chaos in order to Make Sense of the World, *Behavioral and Brain Sciences*, 10.
54, 55

C. A. Skarda and W. J. Freeman (1990), Chaos and the New Science of the Mind, *Concepts in Neuroscience*, 1.
54

P. Smolensky (1986), Information processing in dynamical systems: Foundation of Harmony Theory, in D. E. Rumelhart and J. L. McCelland editors, *Parallel distributed processing*, vol 1, The MIT Press.
76

P. Smolensky (1988), On the proper treatment of connectionism, *Behavioral and Brain Science*, vol 13, pp 407–411.
76

P. Smolensky (1990), Tensor product variable binding and the representation of symbolic structures in connectionist networks, *Artificial Intelligence*, vol 46, pp 159–216.
76

P. Smolensky and G. Legendre (2006), *The Harmonic Mind*, vol 1 and vol 2, The MIT Press.
66, 76, 77, 78, 79, 80, 81, 82, 86, 87

R. Solomonoff (1997), The discovery of algorithmic probability, *JCSS*, vol 55, pp 73–88.
59

K. Stenning and M. van Lambalgen (2007), *Human reasoning and cognitive science*, (to appear).
83, 84, 91

P. Stern and J. Travis (2006), Of Bytes and Brains, *Science*, vol 314, p 75.
73

P. Suppes (2002), *Representation and Invariance of Scientific Structures*, CSLI, Stanford.
3, 42

P. Suppes, D. H. Krantz, R. D. Luce and A. Tversky (1989), *Foundations of Measurement* vol 2, Academic Press, New York, NY.
42

M. Tallerman editor (2005), *Language Origins*, Oxford University Press.
58

A. Tarski (1956), The Concept of Truth in Formalized Languages, in A. Tarski, *Logic, Semantics, Metamathematics*, Oxford University Press, Oxford, UK. (This paper is a translation of the original article Der Wahrheitsbegriff in den formalisierten Sprachen from 1936.)
2, 12

A. Tarski (1959), What is Elementary Geometry, in L. Henkin et al., editors, *The Axiomatic Method*, North-Holland, Amsterdam.
42

L. Tesniér (1958), *Eléments de Syntax Structural*, Klincksieck, Paris.
17

R. Thom (1970), Topologie et Linguistique, in *Essays on Topology*, Springer-Verlag.
17, 52, 81, 83

R. Thom (1973), Langage et Catastrophes, in J. Palis, *Dynamical Systems*, Academic Press.
17, 52, 81

Thomas of Erfurt (1972), *Grammatica Speculativa*, (Edition and translation by G. L. Bursill-Hall), Longman, London, UK.
8, 11, 12

R. Thomason (1974), *Formal Philosophy, Selected Papers of Richard Montague*, Yale University Press.
2, 14, 20

M. Tomasello (1999), *The Cultural Origin of Human Cognition*, Harvard University Press.
72

A. M. Turing (1952), The Chemical Basis of Morphogenesis, *Philos. Trans. R. Soc. London. Ser. B*, London
52, 55

E. Vestre (1987), *Representasjon av direkte spørsmål* (in Norwegian), University of Oslo.
23, 38

L. B. Vosshall (2007), Into the mind of a fly, *Nature*, vol 450, pp 193–197.
68, 92

L. A. White (1947), The Locus of Mathematical Reality: An Anthropological Footnote, *Philosophy of Science Journal*.
40

N. Wiener (1948), *Cybernetics*, The MIT Press.
85

T. J. Wills et al. (2005), Attractor Dynamics in the Hippocampal Representation of the local Environment, *Science*, vol 308.
82

D. Wolpert, K. Doya and M. Kawato (2003), A unifying computational framework for motor control and social interaction, *Phil. Trans Royal Society*, vol 358, pp 593–602.
85